CAMBRIDGE LIBRARY COLLECTION

Books of enduring scholarly value

Technology

The focus of this series is engineering, broadly construed. It covers technological innovation from a range of periods and cultures, but centres on the technological achievements of the industrial era in the West, particularly in the nineteenth century, as understood by their contemporaries. Infrastructure is one major focus, covering the building of railways and canals, bridges and tunnels, land drainage, the laying of submarine cables, and the construction of docks and lighthouses. Other key topics include developments in industrial and manufacturing fields such as mining technology, the production of iron and steel, the use of steam power, and chemical processes such as photography and textile dyes.

Extracts from the Private Letters of the Late Sir William F. Cooke

Originally a maker of wax anatomical models, William Fothergill Cooke (1806–79) became aware of the new electric telegraph while he studied anatomy in Germany. Hoping initially for a return of perhaps a hundred pounds from the English railway companies, he abandoned his studies and turned his attention to the commercial development of the technology, which, though demonstrable in laboratory conditions, was still little understood. Because the process relied on secrecy and many different clockmakers and engineers, it soon became so fraught that Cooke almost gave up before its completion. However, after receiving the encouragement of Michael Faraday and joining forces with Charles Wheatstone, Cooke finally brought his plans to fruition and eventually set up the Electric Telegraph Company in 1846. First published in 1895, this book includes a selection of his private letters, written as he worked and often movingly uncertain, as well as a short memoir.

Extracts from the Private Letters of the Late Sir William F. Cooke

Relating to the Invention and Development of the Electric Telegraph

CAMBRIDGE
UNIVERSITY PRESS

CAMBRIDGE UNIVERSITY PRESS

Cambridge, New York, Melbourne, Madrid, Cape Town,
Singapore, São Paolo, Delhi, Mexico City

Published in the United States of America by Cambridge University Press, New York

www.cambridge.org
Information on this title: www.cambridge.org/9781108052740

This edition first published 1895
This digitally printed version 2013

ISBN 978-1-108-05274-0 Paperback

Willm. F. Cooke

EXTRACTS FROM THE PRIVATE LETTERS

OF THE LATE

Sir William Fothergill Cooke,

1836-39,

RELATING TO

The Invention and Development of the Electric Telegraph;

ALSO,

A MEMOIR BY LATIMER CLARK, Esq., F.R.S.,
Past-President Inst. E.E.

Edited by F. H. WEBB, Secretary Inst. E.F.

The Original Letters are in the possession of the Institution of Electrical Engineers (late the Society of Telegraph Engineers), by whose permission these extracts are now published.

E. & F. N. SPON, LONDON and NEW YORK.

1895.

LONDON:
McCORQUODALE & Co., LIMITED, CARDINGTON STREET, N.W

PREFACE.

THE Council of the Institution of Electrical Engineers have authorised the publication of the following extracts from private letters addressed by the late Sir William Fothergill Cooke to his Mother and to other immediate relatives, believing that they will prove of much interest in connection with the early history of the Electric Telegraph, and its practical introduction into this country by him and the late Sir Charles Wheatstone.

These letters form a very small portion of the valuable and interesting collection of MSS. which have come into the possession of the Institution, and which comprises, besides business correspondence and official papers, a number of letters from men of eminence living at that time.

The collection was originally presented by Sir William Fothergill Cooke and his son-in-law, Colonel Andrewes, to Mr. Latimer Clark, F.R.S., M. Inst. C.E., Past-President of the Institution, and was by him arranged and bound, and presented to the Institution in November, 1891, together with a volume of poems by Sir W. F. Cooke.

The selection of extracts to be published was entrusted to a Committee appointed for that purpose by the Council; the letters having been written at a time when postage was expensive, are almost all crossed, and in some cases even re-crossed.

In order to enable the reader to understand more fully the circumstances in which the letters were written, a short Memoir of Sir William F. Cooke, which was presented to the Institution (then the Society of Telegraph Engineers) by Mr. Latimer Clark in 1879, and which appears in the Eighth Volume of the Proceedings of the Institution, page 361, has been re-printed herewith.

Some difference having arisen between Mr. Cooke and Professor Wheatstone as to their relative positions in regard to the invention and introduction of the Electric Telegraph, they agreed to refer the matter to Sir Marc Isambard Brunel and Professor J. F. Daniell, a copy of whose award will be found on page 93.

It may here be mentioned, as evidence that any soreness of feeling which might once have existed, had been outlived, that Sir William F. Cooke was one of the most genuine mourners who assembled at the funeral of Sir Charles Wheatstone.

The Committee have also added a portrait of Sir William F. Cooke, enlarged by the Swan type process from a *carte-de-visite* photograph given by him to Mr. Latimer Clark; and following the letters is a reproduction in *fac-simile* of a portion of one of them, and of a document expressive of his wish as regards his funeral and burial, the pathetic interest of which is increased by the fact of its being dated little more than a month previous to his decease, which took place at Farnham on June 25th, 1879. He was buried in Farnham Cemetery on June 30th.

COMMITTEE.

G. L. Bristow, Hon. Solicitor Inst. E.E.
Sir Albert J. Leppoc Cappel, K.C.I.E.
Latimer Clark, F.R.S.,
Professor D. E. Hughes, F.R.S.,
W. H. Preece, C.B., F.R.S.,
C. E. Spagnoletti,
} Past-Presidents Inst. E.E.

EXTRACTS.

LETTER I.

HEIDELBERG,

April 5th, 1836.

MY DEAREST MOTHER,

. . . . You must know that for some weeks past I have been deeply engaged in the construction of an instrument which I believe may prove of sufficient importance, should I succeed in bringing it to practical perfection, to merit a visit to London. Determined to satisfy myself on the working of the machinery before I went any further, I prepared to make a model, and, being unable to obtain the requisites at Heidelberg, I sought them at Frankfort. Whilst completing the model of my original plan, others on entirely fresh systems suggested themselves, and I have at length succeeded in combining the *utile* of each; but the mechanism requires a more delicate hand than mine to execute, or, rather, instruments which I do not possess. These I can readily have made for me in London, and by the aid of a lathe I shall be able to adapt the several parts, which I shall have made by different mechanicians for secrecy sake. Should I succeed, it may be the means of putting some hundred pounds in my pocket. As it is a subject on which I was profoundly ignorant till my attention was casually attracted to it the other day, I do not know what others may have done in the same way—this can best be learnt in London.

You see I am very mysterious at present, and think it very prudent to continue so; nevertheless, to you, dearest Mother, if it were your wish, my plan and instrument should be explained now, though I think without better drawings than I could make you would scarcely comprehend me. As I do not wish my motives for visiting London to be generally known, you had better, in mentioning it to my friends at Berne, state that private business requires my presence, and allow them to ascribe to modelling or what they please the sudden change of my plans.

Letter II.

Wednesday, 26th April, 1836.

My Dearest Mother,

I have at length found myself in London once again, after a most agreeable voyage, tho' not with my Heidelberg friends, whom I preceded, they having arrived only this morning. . . . . I have taken lodgings at a pastry-cook's shop in Southampton Place, leading out of Russell Square. . . . .

After commencing this letter, I dined with a Mr. Fergus, M.P., brother of Mrs. Nixon, and there met Mr. Hoppner, who gave me your last letter, which had reached Heidelberg after my departure. Mine from Rotterdam will have answered one leading question—viz., relative to the Doctor's* knowing of my being in England, and the cause. I wish to make no secret of it to him, and were we together I would explain the entire plan, but object to so doing on paper, or its being generally known to our friends, as in case of failure (always a strong probability) remarks, posthumous warnings, and advice are the more overwhelming. I have written to Tom†, begging him to prepare himself with one branch of the subject, and then when I have finished my instrument I will divulge my whole plan to him, and perfect together such papers and statements, &c., &c., as will be necessary ere I proceed further. This mode of proceeding will, I think, be highly satisfactory to you, for, although Tom's knowledge of the world will not aid me materially in bringing my plans or instrument before the public, his clear head and patient investigation will clear away the rubbish from the accumulated heap of facts and statements that I have amassed; and succinctness in my paper statements and calculations will be half the battle.

I have not yet fixed upon my patron (a very important consideration). As the commercial and political worlds are equally concerned, I have a choice between Government and the mercantile potentates. The former are, however, too shy of any

---

* His father, Dr. William Cooke.

† His brother, the Rev. Thomas Fothergill Cooke, M.A.

innovation to listen to any suggestion, however useful, unless strongly supported by higher influence than I can hope to obtain; yet there are several noblemen always ready to parade their patronage when the claimant pretends to originality of design. Whatever course I eventually take will be directed by the best advice I can procure, and then I may hope (it will be all I can do) for a happy result. . . . .

Adieu, dearest Mother.

Ever most affectionately yours,

WILLM. F. COOKE.

In Mrs. Cooke's handwriting is the following note:—

"11th.—Received on the anniversary of his birth, 4th May, 1836. 1st "from his lodgings, Southampton Place, Russel's Square. God preserve and "prosper his undertaking. Replied to before I slept."

LETTER III.

20, SOUTHAMPTON ROW,
2nd June [1836].

MY DEAREST MOTHER,

. . . . Tom will stay with me till I go down to Sudbury, about the 14th of the month—having found a man at Sudbury likely to execute my instrument to my liking. . . . . I heard from the Doctor a few days ago; he wrote me a most delightful and affectionate letter; fully satisfied with my motives for not describing what I am engaged with till we meet. I have explained the instrument and its uses to Robert, who will explain all to you. . . . .

Yours most affectionately,

WM. F. COOKE.

LETTER IV.

2, SOUTHAMPTON ROW,
June 6th, '36.

MY DEAREST MOTHER,

. . . . Dear old Tom, in compliance with my request, came up to dinner on Saturday, looking famously, but well pleased to shake off for a time the monotony of Cambridge. We commenced work almost immediately, and by mid-day yesterday (Monday) he became fully master of the "affair"! Altho' it be impossible to give you even an imperfect idea of "it" in writing, he will give you his idea of its importance and practicability. Robert understands it sufficiently to satisfy you upon the subject, when he arrives at Bern. I have not heard from the Doctor since I last wrote. . . . . It is very doubtful whether I shall be able to go down to Sudbury, as from several causes I am anxious to complete my instrument as quickly as possible, and a simple country workman, however intelligent and devoted to the cause, could only get slowly forward. I fear two months more will pass away ere I can hope to lay plans and instrument before the public; during this time my anxieties will at least keep pace with your curiosity. . . . .

Tom and I are now going to the Adelaide Gallery to study various scientific instruments, connected more or less with our object in hand. His quickness in comprehending all my complexities was delightful; it is impossible to describe the comfort and delight of having him to consult and talk to. My confidence has risen since seven-fold. . . . .

Ever your child,

WM. F. COOKE.

LETTER V.

2, SOUTHAMPTON ROW,
21*st July* [1836].

MY DEAREST MOTHER,

. . . . Old Tom left me yesterday for Cambridge. . . . . His absence I feel terribly; the comfort of having him with me after living quite alone since November last was greater than I can express. He was of great service to me, and very patient under all the explanations I gave and questions I asked. . . . .

22*nd July*. . . . . I think that three months must yet elapse ere I can know the fate of my projects. One instrument will shortly be so far advanced as to enable me to see whether it answers my expectation. I must then have a second made and both finished before laying them before the public. The difficulties you allude to of securing the wires were the first I surmounted before thinking further of the instrument, and having succeeded to my content in that respect I then worked out the remainder. My prospectus is ready, but I am about to send it down to Mr. Chevalier, with detailed drawings, for his judgment and correction. He has seen one movement, and in giving a favourable opinion of [it], offered his further services, little thinking what he entailed upon himself. I will shortly send you my prospectus, but its length will occupy two or three very closely written sheets, and I have not time for that at present. I will only say that, if the wires were broken anywhere between London and Portsmouth, I would find the injury out and repair it in less than eight or ten hours—that is to say, as quickly as a coach could carry me there and back. When you see the prospectus and explanations attached you will be convinced of that. Accident, in the manner in which they are guarded, can hardly injure them; design and malice may. A guard or watch is out of the question entirely. But the mode of discovering the injury when done is both rapid, easy, and decisive. Along railroads the risk from intentional injury is materially diminished, as the depth at which the wires may be laid without increased expense is much greater. Still, occasionally for particular purposes they may be interfered with for a few hours. Still they offer in the long run an advantage. . . . . Still I beg you seriously not to be sanguine of my success. I do not know yet that my instrument will answer, and then very probably it may never be used during my life. I fully believe that the day will come when such a means of conveying intelligence will be employed. I am very far from sanguine, and shall not feel any very bitter disappointment if I fail, provided that my instrument answers the purpose I intend; otherwise, I shall regret having thrown away so much time fruitlessly. . . . .

Your most affectionate Son,

WM. F. COOKE.

Letter VI.

York Chambers, 13, George Street,
Adelphi, London, *Sept. 2nd*, 1836.

My Dearest Mother,

. . . . I do hope this time to have mentioned everything, but I feel far from confident of having succeeded.

I am going with an engineer to look over the St. Catherine Docks, and all the plans, models, &c., which will occupy me till late in the evening. My model is progressing. I hope to see it work before I cross for the Continent. It is now so near completion that I may speak with confidence of its answering its destined purpose. It is a very showy-looking affair. The Doctor was much pleased with it. . . . .

Believe me ever,
Your most affectionate child,
Willm. F. Cooke.

Letter VII.

York Chambers, Adelphi,
Oct. 7th, 1836.

My Dearest Mother,

. . . . My clockmaker has again disappointed me. I called (in Clerkenwell, nearly three miles from here) on Monday evening before coming home, in full expectation of finding that everything had been completed several days, but he told me the balance work had been broken by the running down of the works. This danger I had foreseen, but not obviated, so I gave him a sketch on the instant of a means of preventing a similar accident in future. This, with one or two other little things, was to have been completed on Wednesday. I called—all to be ready on Saturday. This morning a note came to state that the workman who had the balance (very delicate work) to complete had refused to go on with it till he had made something else, but that all should be ready by next Wednesday. Till then I must wait with patience. I then have a good deal of work to fit in myself, and then, if that answers, I have to send it to him again for my index-plate and other things. None but those who have been engaged in such work can imagine the endless series of disappointments and delays to which the inventor of a new and secret instrument is exposed ere he can approach the moment of trial. And then, perhaps the bitterest and ultimate disappointment awaits him. But I am better prepared for disappointment than success. . . . .

Ever your attached child,

WM. F. COOKE.

LETTER VIII.

YORK CHAMBERS,
Saturday, 22*nd Oct.*, 1836.

MY DEAREST MOTHER,

. . . . My instrument was to have been finished this morning, but upon calling I found that a wheel in the escapement movement was wrong, and had to be altered. I am to have it on Monday morning, if not to-night. You would be astonished at the stoicism with which I bear these repeated and endless disappointments. I have been taught a valuable lesson in self-control, if I am to gain nothing more by my labours. The instrument looks beautifully. I hope it will go as well. . . . .

Your most affectionate Son,

WILL. F. COOKE.

LETTER IX.

YORK CHAMBERS,
17*th Nov.*, 1836.

MY DEAREST MOTHER,

Whilst waiting the arrival of my instrument—which *should* make its appearance at six o'clock this evening, but which most likely will not come at all—I will employ my time in writing to you. . . . . I can say nothing certain about my instrument at present. Moore has not behaved well, tho' he has done better the last few days. Tom will have told you that at the moment of his arrival in my rooms I was receiving intimation that one of the most important parts of the machine (the timing part) would not answer. They, however, undertook, by another application of the same escapement, to make it go; but (as I confidently maintained at the instant) they failed entirely. After some weeks' delay, I have since tried another plan, and am to see the effect this night, and in case of its answering I shall be exactly where I expected to find myself on my return from France ! ! ! . . . . My mind is nearly made up to return to modelling if the instrument does not answer; and I suppose Tom has prepared you to expect a thorough discomfiture. I, however, feel myself bound to try all I can before giving it up as a failure. . . . .

Friday. . . . . My instrument came home late last night, but does not answer. I have, however, arranged another plan for keeping the time, which I shall try. Everything depends upon this part of it, as Tom will explain to you. I got the best advice I could from clock and watch makers months ago, and they said they could arrange it to perfection ! ! ! So much for their knowledge. . . . . The moment the probable fate of my instrument is decided, I will let you know. You will perceive, from what I have already said, how humbled are my (never very sanguine) expectations. I shall not give it up till all hope is gone, however. I am very anxious that it should be decided one way before I go to Treeton, as, if unfavourably, I shall make my preparation for renewing my labours in wax immediately on my return. . . . . I am by no means disheartened at this prospect of failure, and shall set to work again, with more than former spirit, at my old occupation. I only regret having talked so much about the Telegraph. The time occupied by this affair, if it should not answer, I do not consider altogether thrown away, as I have acquired a good deal of information which will prove useful one day or other. I also have some other plans in my head, if I can find anyone to take them up, but will not venture on anything else myself unless successful here. . . . .

Ever your attached child,

WILLM. F. COOKE.

Letter X.

York Chambers,
Thursday, 24*th Nov*. [1836].

My Dearest Mother,

. . . . My motive for writing to-day (at a moment when much pressed for time) is to give you a little more cheering news about my instrument than I could do in my last. At that time I had quite despaired of success, and only wished for a justifiable opportunity of giving up, from an impression that I had detected an error in the *principle itself*. I could not at the moment make up my mind to tell you. Tom will understand and explain this. I had good reason for believing that it was a well-ascertained fact that the galvanic fluid only imparted a magnetic quality to cold iron (or an electro-magnet) when its course was short, and that the shorter the course the greater the attractive power. Hence electro-magnets were made with several short thick wires for experiments requiring *quantity*, and one long thin wire where *intensity* was required! To set this point at rest I got an introduction (through Dr. Uwins) to Faraday, the "King of Electro-Magneticians." He received me yesterday, and I asked him to give me his opinion upon the instrument, and in the kindest manner he proposed calling this morning for half an hour, which he did, but stayed an hour and a quarter, entering with great interest into all the details. He finally gave me as his opinion that the "*principle was perfectly correct*," and seemed to think the instrument capable, when well finished, of answering the intended purpose. He would not give an opinion as to the distance to which the fluid might be passed in sufficient quantity, but observed that if it be only for 12 or 20 miles it can be passed on again. He said in reply to my question: "I am "afraid of inducing you by my advice to expend any large sum "in experimenting, but it would be well worth working out, and a "beautiful thing to carry on in this manner a conversation from "distant points; and the instrument appears perfectly adapted to "its intended uses."

Now I consider this highly satisfactory. He took his leave in a most friendly manner, but in a way which induces me to think he does not mean to take any further step in the affair. I asked his advice as to my way of proceeding in bringing it before the public when completed, but he declared his inability to advise me. I do not intend to expend anything more upon it myself, but hope to find someone who will take the risk in consideration of a fair remuneration in case of success. The difficulty of arranging the escapement I shall be able to overcome, beyond a doubt. You will see by what I have said that if I fail it will not be owing to any obvious error, which I ought to have detected myself; and that is a great relief to my mind, for I have felt the disgrace of failing from ignorance as to the principles of the power I employed, which would justly have exposed me to severe animadversion for presumption.

I showed him also my *perpetuum mobile* principle. He says it is original, to the best of his belief, but deemed it difficult to work out. He was so pressed for time that he could not enter into the nature of the means I proposed employing to attain my object. This coincidence with Mr. Chevalier's opinion is highly satisfactory, even should it come to nothing. . . . .

I still think of going on with my modelling, not to be idle, and to put a few pounds in my pocket if I can. . . . . Do not let your hope of my future success be too sanguine, but that there is a fair ground for the further prosecution of the plan I consider proven. . . . . Whatever be the ultimate fate of my plans, I shall certainly derive some benefit from the moral tuition I have undergone, and have, I hope, acquired a greater degree of control over myself than I before possessed.

With most affectionate love to the Doctor and Tom,

Ever your attached child,

WILLM. F. COOKE.

P.S.—Though I have been up till 2 o'clock on Monday and Tuesday, and 4 o'clock last night, I have no headache.

Letter XI.

York Chambers,
28th November, 1836.

My Dearest Mother,

. . . . I perfectly agree with your observations respecting Faraday. I feel a great satisfaction in knowing that my views on the subject and the principle of the instrument are all right. My undertaking it is thereby justified.

I have given Moore (who is the bearer of this) my new plan for the escapement. He acknowledges its superiority to our former plans, as its resistance can be modified to any extent. Faraday could not speak to the distance that the fluid can be conveyed, as he does not know himself. Experience alone can prove this. The most minute quantities have been conveyed 4½ miles (English measure) without any perceptible abatement of their force. The Germans are of opinion that a thousand miles would make no difference. Faraday says that remains to be proved, but does not deny the position. I am still of opinion that my only way to ensure success is to part with some of the advantages to be derived from it to those who are to bring it forward, and, as I am entirely unknown, I must get what aid I can, being unable to chuse [*sic*].

You are right, dearest Mother, not to be too sanguine. The projectors of an original undertaking hardly ever yet reaped the benefit of it. The Steam Engine, Gas, &c., ruined the inventors. The E.M. Telegraph shall not ruin me, but will hardly make my fortune. If it make me friends I may get on well enough. . . . .

Believe me, my dearly loved Mother,
Your warmly attached Son,
Willm. F. Cooke.

Letter XII.

York Chambers,
December 8th, 1836.

My Dear Father,

You will be surprised to see my handwriting again so soon, but an idea has crossed my mind which seems to brighten my hopes after a morning's despondency, and as you can perhaps aid me I will not lose a post in imparting it to you. Two applications to scientific and influential personages have been made for me without success, and I have heard nothing more from Faraday. In this dilemma I thought of Mr. Walker, whose tastes, connections, situation in the neighbourhood of rail-road and speculating influence, and friendship for you render it possible that he may interest himself about my Telegraph. I think I have heard you say that he has entered into some of the speculations in the neighbourhood of Liverpool, and may therefore not object to trying another where the expense will be very trivial, with still less risk, and success must be attended by honour and emolument. I am willing also, to give up a considerable proportion of the advantages to any-one who will forward my objects—taking out a patent and trying the necessary experiments after completing a second instrument. If you think Mr. Walker a likely person, and would write to him on the subject, you may state any of the circumstances connected with my plans that you please. My invention has a national and commercial object in view. The instrument and principle, coming peculiarly under Faraday's department, have been approved by him. I am perfectly willing to show the instrument and explain all my views to Mr. Walker, as a preliminary step, before any conditions are entered upon, and obtain (if possible) a written opinion of them from Dr. Faraday. My wish would then be, if Mr. Walker approve, and be inclined to proceed with me in experiments, that he, and any friends of his who would unite with us, furnish the means of proceeding with the patent and instruments, receiving in return shares. If you please, you may mention the nature of the instrument, in which case you had better add that methods of securing the wires and of detecting injury done to them, have been carefully arranged to Faraday's satisfaction,—that the instrument I have made is a "working" one of full dimensions. The expenses attending patent, another instrument, and experiments will not, I imagine, exceed three hundred pounds (£300), even if tried on an extensive scale of many miles. My object for applying to Mr. Walker being to find a person or persons connected with the busy world who could effectually bring my instrument and plans before the public, and would join me in the expense and profits. I feel pretty certain that, could I once get connected with such men, I have other means of pushing my own fortunes, and am therefore the more willing to make a considerable sacrifice at the outset. If you do write to Mr. Walker, let him fully understand that he binds himself to nothing in agreeing to see my instrument but "silence;" and as I am *inclined* to pay the North of England a visit very shortly, I should not even be put to the expense of a journey expressly for that purpose. You might add that I am anxious to get forward, and would therefore consider an early reply to your letter as a particular favour. Under existing circumstances I should be too happy to find employment in the neighbourhood of Rotherham! Mr. Walker would, it appears to me, be a delightful person to become acquainted with in such speculations, and I should sincerely rejoice to have in him a warm co-operator. I must now acknowledge that dearest Mother, some months ago, mentioned Mr. Walker to me (at least I think so) at Paris, but at the time my views were different from what they are now. . . . . I leave you at perfect liberty to act as you please; and if you think it desirable to explain more fully my system and object, Tom can furnish you with a clearer statement than I could. Perhaps a little confidence would interest Mr. Walker the more.

The greatest distance the galvanic fluid has yet been conveyed for experiment is 4½ English miles; two very small plates were used for the battery, and no diminution of the activity of the fluid was observable. *I* use powerful batteries. . . . .

Ever your most affectionate Son,

WILLM. F. COOKE.

LETTER XIII.

YORK CHAMBERS,
December 16*th*, 1836.

MY DEAR FATHER,

I am most thankful to you for the prompt manner in which you complied with my wishes in writing to Mr. Walker, and will not fail to let you know his reply as soon as I receive it. If it be favourable, I shall take my instrument with me to the North, and should he or any friends be inclined to take it up I shall feel that I am in good hands. It is most probable in that case, the further prosecution of the experiments will be nearly taken out of my hands and placed in those of scientific and practical men—such, at least, I should both advise and wish. In which case, till another instrument is made, I shall have my time unemployed, and propose going to work at modelling upon my return from the country. . . . [The rest of the foregoing letter is devoted to the explanation of a scheme for the formation of an anatomical museum.]

Ever your affectionate Son,
WM. F. COOKE.

LETTER XIV.

[27*th February*, 1837.]

MY DEAREST MOTHER,

. . . . I had hoped to-day to have been able to give you some certain intelligence of my own plans, but I am unable to fix upon anything. I tried last week an experiment on a mile of wire, but the result was not sufficiently satisfactory to admit of my acting upon it. I had to lay out this enormous length of 1,700 yards in Burton Lane's small office in such a manner as to prevent any one part touching another; the patience required and fatigue undergone in making this arrangement were far from trivial. From Monday evening till Thursday night I was incessantly employed, and by Friday morning at 10 o'clock all was obliged to be removed. Dissatisfied at the results, I this morning obtained Dr. Roget's opinion, which was favourable, but uncertain; next, Dr. Faraday's, who, tho' speaking positively as to the general results, formerly, hesitated to give an opinion as to the galvanic fluid action on a voltaic magnet at a great distance, when the question was put to him in that shape. I next tried Clark, a practical mechanician, who spoke positively in favour of my views; yet I felt less satisfied than ever, and called upon a Mr. Wheatstone, Professor of Chemistry at the London University, and repeated my query. Imagine my satisfaction at hearing from him that he had four miles of wire in readiness! and imagine my dismay on hearing afterwards that he had been employed for months in the construction of a Telegraph, and had actually invented two or three, with the view of bringing them into practical use. We had a long conference, and I am to see his arrangement of wires to-morrow morning, and we are to converse upon the project of uniting our plans and following them out together; from what passed, my plan, if practicable, will, I think, have advantages over any of his; but this remains to be proved. Under all circumstances, I should be happy to have a scientific man for my coadjutor, though in that case I must sacrifice a larger portion of the advantages—yet I value his aid much more. Tom will know him as the person who invented the plan for ascertaining the velocity of lightning by means of a revolving mirror. But for the fear of increasing your anxiety at dear Madge's silence I should not have written till after our meeting to-morrow, though most likely several days must yet elapse ere anything definite can be concluded. I cannot say I have enjoyed myself much since coming to town. Every moment has been anxiously occupied. I have been engaged in moving rapidly from place to place and holding these interviews ever since ten o'clock, and it was four when I commenced writing. I do hope a few days will enable me to decide upon some plan, and either condemn my instrument or set it going. This suspense becomes too wearing. . . . . I shall be very cautious in employing the money the Doctor has so liberally placed at my command in taking out a patent, unless I see my safety very clearly, which I should do if acting in conjunction with Mr. Wheatstone. . . . . At present all is doubt and uncertainty

—in truth, I had given the Telegraph up since Thursday evening, and only sought proofs of my being right to do so, ere announcing it to you. This day's enquiries partly revive my hopes, but I am far from sanguine. The scientific men know little or nothing absolute on the subject. Wheatstone is the only man near the mark. I cannot explain the point on which so much depends, to you on paper—Tom may—when I say that a lengthened course *may* convert *quantity* into *intensity*, in which case no magnetic quality is imparted to the iron magnet, as *quantity* alone produces that effect. This is a very subtle point, on which the "Doctors" know nothing, and had not thought till I put it into their heads. . . . . I shall not regret all this in the end if I succeed at last, as meeting and conversing with all these men will do me good in many ways, and I make nothing now of composing my mind, after having reason to believe that all my labours have been in vain. . . . . Wheatstone is a music-seller in Conduit Street! but an extraordinary fellow. . . . .

With affectionate love to old Tom,
I am ever,
Your attached child,
WM. F. COOKE.

LETTER XV.

19, BURY STREET, ST. JAMES,
14*th March*, '37.

MY DEAREST MOTHER,

I have not had a moment to spare till to-day for letter writing even to Treeton, so must be excused for not giving you earlier notice of the unsatisfactory results of our experiments. Mr. Wheatstone called on Monday evening, and postponed our meeting at King's College till Wednesday. The result was nearly what I had anticipated, the electric fluid losing its magnetising quality in a lengthened course. An idea, however, suggested itself to Mr. W. which I prepared to experiment upon last Saturday, but again failed in producing any effect.

I gave up my object for the time, and proposed explaining the nature of my discomfited instrument to the Professor; he in return imparted his to me. He handsomely acknowledged the advantages of mine, had it acted. His are ingenious, but not practicable. His favourite is the same as mine made at Heidelberg, and now in one of my boxes at Berne, requiring six wires, and a very delicate arrangement. He proposed that we should meet again next Saturday and make further experiments. For a time I felt relieved at having decided the fate of my own plan, but my mind returned to the subject with more perseverance than ever, and before three o'clock the next morning I had rearranged my unfortunate machine under a new shape. Tom will understand me when I say that I now use a true magnet of considerable power, with the poles about 4 inches apart, and suspend on the plane with its poles by a pivot (like a mariner's compass) a slender armature 4½ inches long, covered with several hundred coils of copper wire (covered with silk). On passing through this coil a stream of galvanism the armature becomes polarised, and is attracted by the poles of the magnet. The magnet, formed of a cylindrical rod of cast steel, curved, is thus shaped: N S. The armature is represented by the dots, one end being on this side of the north pole, its fellow on the other side of the south pole, seen from above, thus:

the ends of the armature magnetised, Z negatively and C positively, whenever the galvanic circuit is completed, and then respectively attracted by the N. and S. poles of the magnet with a sufficient force to enable them to overcome the opposition of a feeble spring, the movement will not exceed

1-20th of an inch; a lever, forming part of the detent of my fan, is moved by the pin projecting from P, and liberates the clockwork. Tom has seen an arrangement of this description in the Adelaide Gallery, used there merely as [a] toy.

I have determined upon making two instruments on this principle, and trying them with Mr. W.'s four miles of wire. Moore has already commenced them. I have made further simplifications of the plan I showed you at Hastings, using no quicksilver, triangles or springs, every part restoring itself, by balance weights. The cost of two will not exceed, I hope, £8, and it [*sic*] may be made hereafter for about £2 each I hope. If I can reduce the cost to this amount, I hope to be able to introduce it into our public offices, banks, &c., and perhaps private houses, as I have a battery in view, about *four inches* cube, which will continue in action for 10 days or more without cleaning, no acid used. I may also get it employed in short distances on railroads at once, and experiment may enable us to use it for greater distances hereafter, when the requisite thickness of wire is better understood and batteries in greater perfection; thus you see I am deeper in it than ever. I am hastening to a philosophical coterie this evening, where I hope to gain further insight into the construction of the galvanic arrangements.

Your attached child,

WILLM. F. COOKE.

LETTER XVI.

19, BURY STREET
[8th *April*, 1837].

MY DEAREST MOTHER,

. . . . I have been extremely busy with enquiries and experiments on the galv[c] battery, and am highly pleased with the result. Tom shall have them in a future letter. I have hit upon one important point, and formed a battery upon principles likely to secure my object; of this more hereafter. I commenced at 7 yesterday morning, and never left the room till 2 this morning, working without intermission for 19 hours.

This exceeds what I ever did when modelling. My instruments are going on famously, though somewhat slowly. I think they must answer. . . . . Mr. Walker's enquiries at Liverpool were unsuccessful. His letter contained kind offers of service at any time. My plan of proceeding when the instrument is finished is not yet finally arranged.

With kindest love to the doctor and old Tom,

Ever your most affectionate Son,

WILLM. F. COOKE.

LETTER XVII.

16, COMPTON STREET,
BRUNSWICK SQUARE,
April 13*th*, '37.

MY DEAREST MOTHER,

. . . . I have left Bury Street, being in need of an extra room to try galvanic experiments, which I am conducting on a varied scale, with all the minuteness, accuracy, and ingenuity I can summon to my aid. The result will, I flatter myself, be the most efficient battery for experimental or telegraphing purpose yet brought to light, and by far the cheapest. I can keep up perfect steadiness of action for a great length of time, and the moment I have done with the battery a touch of the hand removes it from the action of the acid, and by exposing the plates to the air restore them to their original activity. None of that destructive local action goes on, so fatal to the ordinary battery, and disagreeable from the quantity of gas evolved, all action ceasing the moment the galvanic circuit is completed. From the simplicity of the construction of my battery, anybody can make one at home equal to the best that can be bought, and restore a wasted rod of zinc in the course of half a minute. At present, I see little more to desire in this part of my apparatus. My instruments will both be finished early next week. Moore is making up for past neglects. . . . .

Ever your affectionate Son,

WILL. F COOKE.

LETTER XVIII.

[*25th April*, 1837.]

MY DEAREST MOTHER,

. . . . My instruments are not yet finished, but progressing rapidly. I hope my next will inform you of ulterior proceedings with regard to the patent. . . . . All goes on most promisingly, particularly my galvanic arrangements, which promise great things. . . . .

Ever your attached child,
WM. F. COOKE.

LETTER XIX.

16, COMPTON STREET,
BRUNSWICK SQUARE,
3rd May, '37.

MY DEAREST MOTHER,

. . . . I do congratulate myself sincerely on being able to give you something like good news respecting my instruments, both of which I have had at home and in working order

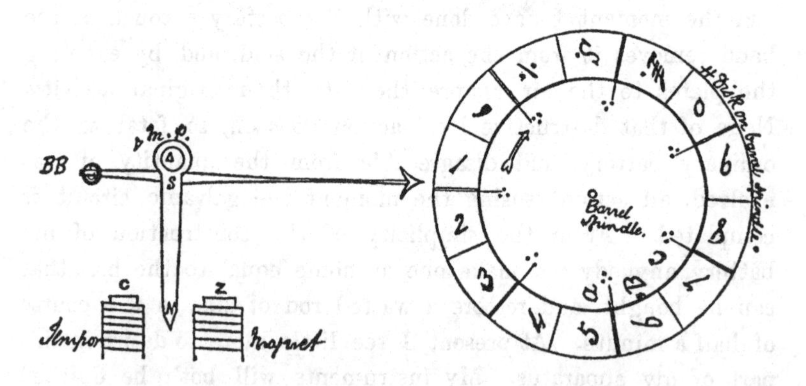

since Saturday. By a better and more comprehensible method than that I explained in my letter from or after my return from Liverpool I obtain 41 signals; and, as my head is too full of my plans to allow of my writing rationally on any other subject, I may as well give a slight sketch of my plan, as it is now arranged, which Tom will further illustrate.

My barrel has seven pins, corresponding with as many moveable keys, each giving me a distinct signal. By this means I can represent numbers in reference to the signal book to that amount. On the barrel spindle I fix a disk of metal, as in my original instrument, which revolves with it, with the following numbers, &c., inscribed upon it. The seven long double lines are the points corresponding with the seven pins of the barrel and seven keys, which alone, by stopping opposite to a fixed hand, would only furnish me with as many signals. To double the number, I suspend between the legs of the temporary magnet (which is the prime mover of the instrument) a small steel magnet, on a very free pivot, attached to which is a delicate index or pointer, with a small *balance ball* on the short arm. All this is on the outer face of the instrument, and will have a glass fixed over the index-plate, or disk, so as to show it and the arrow-heads of the index, as in a common clock. The rest of the instrument will be concealed in its wooden frame. When one instrument of a pair is worked, the check of both is withdrawn by the temporary magnets, by means not seen in the accompanying diagram, and the barrels continue to revolve till the key which has been set is struck up by the corresponding pin in the barrel. The current of g. fluid is thus interrupted, and, the temporary magnets losing their magnetic character, the check falls and stops both instruments simultaneously. In the diagram, 1 or 2 would be indicated, and the arrangement of the index decides which. The index magnet has its north pole downward, consequently when the leg Z of the temporary magnet is under the influence of the zinc (negative) end of the battery it attracts the temporary magnet, which remains (by its own attractive force to the cold iron) inclined towards it after the current is interrupted, and the index point shows 1; to reverse this action the current is to be sent in a contrary direction, and 2 would be indicated; that the operator may more distinctly perceive what he is about, each key will have its corresponding numbers engraved on it, the one on a black

surface, the other on a white. The commutator by whicl
are changed will have two pieces of ivory attached, one
one white. When a white figure is to be shown, the

the commutator will be moved to the left, when the black ivory will sink below the frame and the white one will rise above. The operator then proceeds to deliver his notices as long as the numbers to be represented are "white." When a black one is needed he changes the direction of his commutator, and then touches the corresponding key, &c. Supposing signal book numbers to be represented, I get then by this arrangement: 1, 2, 3, 4, 5, 6, 7, 8, 9, 0, , —total, 11 in reference thereto. As by my old plan, the signal 0, signifies that what follows are figures of amount, and I get 1, 2, &c., as before, up to 11, and £, s, ÷; total, 14 (the last my fractional sign). Now I do not like to lose the power of spelling, and I obtain it in this manner. In the interior of my numerical index I place letters, of which A, B, C are given in the diagram, and on the top of the frame through which my key heads move I write the corresponding letters; £,, in the numerical plate, if not for my figures, which would signify so many pounds, stands no bad type for "*Vide* "letter," and the index point which previously gave 5 will now mean A, as 6 will be supplanted by B, 7 by C, &c. Now these will only give me 13 letters, and the necessary stop (,), which I must make do for the entire alphabet, as I object to divide the time of my barrel's revolution by more pins and keys. This I effect in the following manner, which, being hastily arranged the other night, is doubtless susceptible of great improvement. A, H, L, O, R, S, , will represent themselves, and "no mistake." B, C, D, will represent perfectly P, K, T, being by the Welsh and other Mahomedans substituted for them, as Packgammon, Kat, and Citten, Tick for our familiar Dick, &c.

A
B = P
C = K
D = T
F = G
H
I = Y, U, V, W
L
M = N
O
R
S
X = "Exit," if followed by ,
,

M and N may occasion a mistake if the context of the sentence did not explain, as in Moon, Noon, Sane, Dine, Din, Dome, and the French Ton for my brother Tom, both being sometimes "Bon ton" or "Good Tom." I for Y, U, V, W, can hardly admit of error; and X followed by a stop or alone signifies prettily "Finale," and permission to return to the figure-row, when £ , signified amount or numbers by their appearing. E, I omit, as all the letters are sounded with it except K (Kay), R, and W, and E is written in no Eastern language. J, Q, Z, I have nothing to say to. We have thus accounted for 11 numbers, 14 in figures, 14 signals in letters—total, 39. I can add two more by second meanings to S, ÷

Now, my dearest Mother, I have written thus much for Tom's edification, rather, I fear, than for your amusement. Before concluding this subject, I will add that my instruments in every respect come up to my most sanguine expectations, working beautifully. There may be, however, difficulties yet to overcome which I do not now foresee. Professor Wheatstone is anxious to secure the co-operation of an influential house in the City, and till this point is settled no patent will, I fear, be gone on with. I must have patience, therefore, for a few days more. . . . . I wish I had old Tom here to work out some of my papers and go on with my experiments; together they are more than I can attend to. . . . . I hope by this day week to have commenced with more decisive operations, and when the step is taken you shall know. Three weeks are required for the passing of a patent through all its stages. . . . .

With affectionate regards to the Doctor and old Tom,

Ever your warmly attached child,

WILLM. F. COOKE.

LETTER XX.

16, COMPTON STREET,
BRUNSWICK SQUARE,
Friday [*4th May,* 1837].

MY DEAREST MOTHER,

I write these few hasty lines to say that it is just possible that I may be able to start for Hastings by to-morrow's mail. . . . . I must certainly be back by Wednesday. My leaving town to-morrow depends upon the result of a meeting I am to have with Professor Wheatstone at 12 o'clock. If I do not get off in the evening you must not expect me till the following Saturday.

Yours affectionately,

WM. F. COOKE.

I had a conversation thro' the two Telegraphs with you yesterday.

You began, " My old Boy."

" Yes, Mother."

" 4th May."

" Success."

" Persevere."

And I had just time after this bit of play to dress, and then off to the Lanes'.

Letter XXI.

16, Compton Street,
Brunswick Square,
11*th May* [1837].

My Dearest Mother,

. . . . At King's College I was introduced to the Messrs. Enderby, who rank among the leading men in their line in London—enterprising, determined men. They were much pleased with the simple experiments shown them, and say there will be no difficulty in raising the capital, but did not offer to do it themselves—of course they must see our instruments first. . . . . As I may hereafter find it convenient to have my plans and improvements established by letters bearing post-mark and date, I shall describe what I did, for Tom's immediate satisfaction and my own ultimate security.

The changes about to be described refer to the instrument made at Heidelberg on the old principle—not the chronometrical arrangement. They are of so promising a nature that the old discarded plan will probably take the lead in some cases and situations, after all.

To Tom: You understood, I think, the arrangement by which I made four wires complete three circuits, and, by reversing the same, three more. It may yet be well to describe.

Suppose, then, two batteries with four lines of wire. Now, by uniting the wires at Z, C, when both batteries are disconnected with them, a current may be passed from C′ (the positive pole of the battery) through any of the lines 1, 2, 3, to the negative pole Z′. Reverse the poles, and the positive current passing along wire 4, may be made to return by wires 1, 2, 3. By this means you obtain six simple signs; when the communication is concluded, the wires 1, 2, 3, 4, are all united, and remain ready for receiving signals from the other end of the line. By pausing double time on any of the contacts, a second signal may be represented by each wire, making the number of signals twelve. By connecting two lines of wire with one pole a fresh variety of signals may be given, exceeding in number those required for telegraphic purposes.

Now my present arrangement of the old instrument is this:

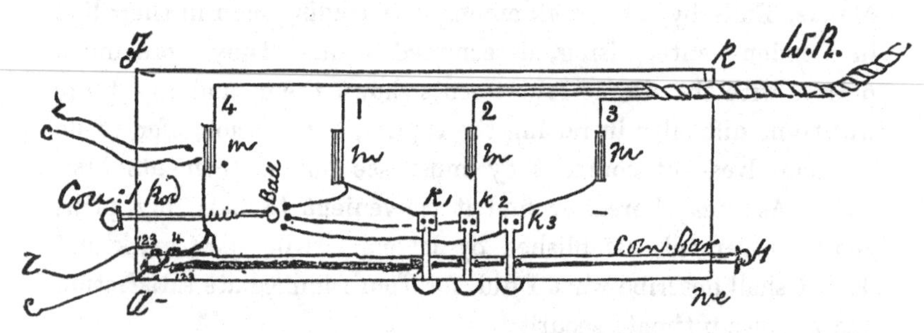

W R, Wheatstone's wire rope, consisting of four wires, each distinctly covered with tarred flax and then included in a rope. This is an idea of mine which will answer well, if the counter currents passing for a great distance in close proximity do not interfere with each other—1, 2, 3, 4, the insulated wires, forming as many multiplicators, and then proceeding to their respective keys, which will have their heads firmly screwed into the frame and extend their stiff springy arms beyond the border of the F, R, A, *me*, and terminate in bits of ivory. Each key has a springy # piece of brass projecting below the arm,

thus, when the key is strongly pressed down, that springy bit of brass which forms part of the conducting wire, comes into forcible contact with the commutator bar. This bar is of wood, with two separate strips of copper attached to it, cut thus:

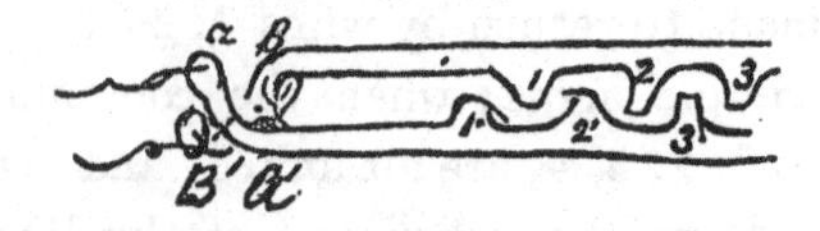

The bar slides to and fro about an inch, being pushed by the H (handle). You will conceive, then, that when the key 1 is pressed down its springy bit of brass comes forcibly in contact with the point 1 in the upper strip, as also keys 2 and 3 with their corresponding portions of the copper strip, but if the commutator rod is pulled out, the keys come in contact with their corresponding portions of the under copper strip, and this copper is so bent as to bring the points 1′,1, 2′,2, 3′,3, in a line with each other, favourable for the key to play upon. On the finger being removed the key recovers itself, and the contact is broken. At the left extremity of the com. rod the copper strips are bent in contrary directions round the rod, and offer four points for contact, against which the springy ends of the battery wires, Z, C, press at one time Z on A, making the lower copper strip positive—but no—All wrong—This would bring me back to where I was before, so away with the sliding rod and double copper strips, and I will only move my battery poles. Look, then, to this drawing, and you will observe the fourth wire passing to the connecting rod and thence to the knobs 4, 4′, which are now in contact with the C pole—the Z pole communicating by the other knobs, marked 1, 2, 3, with the commutator rod ready for contact with the keys. By advancing the pole wires the circuit will be reversed. When the connecting rod is advanced the ball comes in contact with the three balls of the wires 1, 2, 3, and the instrument is ready to receive impressions from the other terminus. On being withdrawn it is prepared to become the speaker. The 4th multiplicator is the discharger of a home battery to sound the alarum, which I have still further simplified. Above the other three multiplicators hang, in tubes of glass, three silken threads supporting their needles, with pith balls, having their respective signs as before described to you. [All this is] concealed in a wooden frame, except the three keys and three glass tubes,—the instrument will appear thus:

The glass tubes will only be 4 or 6 inches in height by 1 inch diameter, and the whole affair very small and cheap, the four wires being the objection, which will be diminished when the distance is short. I should have said that the handle that changes the poles will also move an ivory slide on which the characters represented are engraved as before, half on black and half on white, so that the letters connected with each polar movement will slide in front of the key when they are to be represented, and in the same colour as they appear on the pith balls; this will preclude confusion, even on the first attempt.

Do you understand this wildly confused description? I have not had time to make out my drawings yet, and you know how difficult it is to carry those polar changes in one's head; but I have got the thing pretty clearly now. By this plan I can readily give more than 60 signals!!! Indeed, there is no bound; but 68 would be easily recognised, and 36 all simple signs!!!

LETTER XXII.

Long details for Tom.

16, COMPTON STREET,
BRUNSWICK SQUARE
[23*rd May*, 1837].

MY DEAREST MOTHER,

I write, as I think you will be expecting to hear from me, in reply to your two notes, but have nothing particularly interesting to write about. Every moment is busily occupied in making arrangements and preparing instruments for our experimental day. This holiday week past has thrown our patent back, which cannot be out before the end of the month, if so soon, and consequently no exertion of mine can place things in such a state of forwardness as to enable me to accompany the Doctor on the 31st May or 1st June to Treeton. . . .

I heard from Mr. C. Enderby yesterday that Mr. Joshua Walker seemed favourable to the undertaking, and that Mr. E. thought he would embark in it. If so, we must give him good terms, as he will enter with strong interests in London and very great at Liverpool. The Enderbys are a very wealthy family, who have made most of their money in the South Sea trade. They have a large sail-cloth and rope manufactory at Greenwich, and a great many other undertakings on hand. They appear very friendly to me. I am going down to Greenwich to-morrow morning to see their works, and arrange a method of covering our wires with yarn and include them in a rope, for our cross-Thames experiment. I sent the wire off yesterday. This rope, 1,500 feet in length, including 6,000 feet of wire, is to be ready by the close of the week.

My two new instruments on the Heidelberg plan are in a great state of forwardness. Wednesday will see one finished, and the other nearly so. I have the frames and keys made, and made some of the multiplicators last night. They will all be finished to-morrow before starting for Greenwich.

I am making a model for [? of] Prof. Wheatstone's instrument.

After seeing Mr. Jos. Walker next week, I shall, if encouraged by him, proceed with a rope and wires for the London terminus of the Birmingham Railroad, where they require a signal communication. Hoppner furnishes me with the exact measurement. The same rope I take with me—or, rather, send by sea—to Liverpool, to exhibit there. But as we do not intend risking too much ourselves, we shall rather look to Mr. J. Walker for authority to do this on his responsibility. With so many things to attend to, and different workmen to look after repeatedly during the day, I have not leisure to think quietly on any subject. . . . .

[Then follows a description of his plan for electric communication between fire brigade stations, the wires being enclosed in street pipes, and signalmen being stationed on church steeples and other high buildings. He requests "Tom" to prepare a statement of his plan for submission to the Government and the Insurance Companies.]

I have rendered my Heidelberg instrument more compact, and reduced my wires to three, after much thought, by a plan which might have struck me (you would have thought) sooner. Fancy the three tubes, A, B, C, with their dependent signals as described in my last letter, and with only *two* keys. Three wires are in connection with the multiplicators below each tube, the current passing up the wire of A or B accordingly as the key

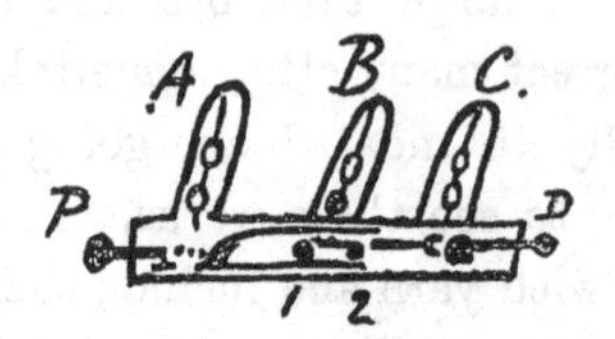

1 or 2 is pressed, and returning by C. Consequently two needles —viz., A and C, or B and C—must always move at each signal. C is disregarded, and the signals read off from A or B, which, as before, would give eight signals; but if both keys are pressed at the same moment, all three needles turn as the divided current travels up A, B, and returns by C. . . . . We can make nothing out of our old friends A, B. We at last condescend to notice the ever-active C, who furnishes us with four signals more. My pole commutator I have borrowed from the little battery I took with me to Hastings last time; it is worked by pulling out or pushing in the knob P. The dots represent the insulating ivory which separates them. In connection with the same, and passing above the two keys, may be seen a line representing a piece of ivory, with the characters to be given inscribed upon it, and appearing under glass at the top of the frame. The glass is blackened in alternate stripes, so that when the knob is pulled out the black stripe conceals those numbers which cannot then be given. On pushing in the knob the poles are changed, and accordingly the numbers which could not previously be seen become visible, and upon pressing the key corresponding with either of them the needles turn so as to show the same signal.

My keys are of brass wire, when at repose pressing against a bar of brass in connection with the wire C, thus placing the instrument in an attitude of "attention." When speaking, the knob D is drawn back, which breaks the connection with a receiving cup, and the keys being pressed down firmly on a slip of brass forming part of the pole commutator, the battery current becomes complete.

"Do you twig?"

I am afraid this instrument will supersede the chronometer in most cases. If so, how much labour has been thrown away! . . . .

I have just returned from Greenwich, and have given orders for the rope on a far cheaper plan than was previously proposed. Its cost will be a mere trifle. Adieu; just in time for the post.

Yours affectionately,
WILLM. F. COOKE.

LETTER XXIII.

10*th June*, 1837.

MY DEAREST MOTHER,

. . . . Enderbys Brothers write that they have finished the rope. At 4 yesterday I went to King's College to meet Professor Wheatstone and try my instruments, which have nearly received their last touch. . . . . I had hoped to have had our experiments made public to-day, but dare not till the patent is out, as one day's impatience may ruin all. The King's health is still so precarious that he can transact no business. A report was very prevalent yesterday that he was dead, but contradicted in the even^g.

Again adieu; with every aff^tc wish for many and many a happy return of this day, believe me, my own dear Mither, your attached and grateful child,

WM. F. COOKE.

P.S.—Hurrah for the 10th of June.

On the back I send you good news, this moment (5 minutes to 10 o'clock, 10th June) obtained. All now is safe. In haste.

[On the back of the letter is written as follows:—]

Cooke's and Wheatstone's Patent signed by His Majesty, and receiving the Great Seal this day, June 10th!!! 1837, for Electric Telegraph Alarums.

I had intended to send you some small present, but now send this instead.

Letter XXIV.

16, Compton Street,
Brunswick Square,
2nd July, 1837.

My Dearest Mother,

Tho' having good news to impart, I could not find time till now to write you a few lines, every moment being engaged from 6 in the morning till 10 or 11 at night. On Friday, the day after I saw you off with Charles, I called on Mr. Joshua Walker, and imparted my plan for the fire telegraph. He spoke handsomely of it, but recommended my proving the practicability of the general principles before I attempted to introduce a project involving the disturbance of the pavement. I then expressed my wish to try experiments on the railroad. "There," he said, "I can at once assist you," and within half an hour introduced me to the Chairman and Secretary of the London and Birmingham Railroad. They both entered warmly into my views, and appointed the following day for a further consideration of the subject. To shorten details, by following up every opportunity that offered itself, and urging forward my suit unceasingly, I got through all the forms, had three interviews with Mr. Stevenson, the famed engineer, and got an order for eight hundredweight of copper wire, by Friday last; obtained leave to occupy a vast building on the railroad, 165 feet by 100 wide, and had as many men and all the materials I could require placed at my disposal. The order was, "Let Mr. Cooke have everything "he may require."

By strenuous exertions I succeeded in collecting the above vast quantity of wire, cleared the huge workshop of men and lumber, by the constant labour of from 30 to 40 men, and had nearly half a mile of wire arranged by Friday night; proceeding slowly on Saturday morning, having to teach all the men employed—viz., 8 carpenters, 2 wire-workers, and 8 boys—their distinct duties, we got forward more rapidly towards evening, and at 5 o'clock, when the men left off work, I had about four miles of wire well arranged, and hope to get all nearly done by to-morrow night. You may imagine the task when I tell you that 2,888 nails have been put up for the suspension of the wires. The labour can only be conceived by witnessing our proceedings. I am anxious to show as much activity and accuracy in the execution of my arrangements as possible, or I might proceed more leisurely, and in any manner I pleased. The Secretary, however, enquired from the Chairman whether I would superintend the laying down of such a communication along the line to Birmingham, should the Company hereafter determine upon it; and I am very desirous of proving my ability to undertake such a task. Should my movements progress as I could wish, I hope to show my experiments to Mr. Stevenson on Thursday or Friday, when he will be in town. He having been appointed by the Directors to report thereon.

My application in the first place was to try an experiment at my own expense, but, finding the parties I addressed so ready to listen to me, I finally proposed their bearing me through free of expense, which they unhesitatingly have done in the most liberal manner, at an expense of about £60. . . . . I am not over-sanguine, but congratulate myself on this opportunity of having fair play and a public trial.

I had a most affectionate letter from the Doctor yesterday, making me a most noble offer of assistance, to an extent nothing would have induced me to accept, even had I needed it. I rejoice, however, to receive such a noble proof of his affection, and, in reply, to be able to satisfy his mind that I had obtained the desired object without risking further capital.

These Companies are *slow coaches* to drive, and months may elapse ere they take any further step, unless constant applications urge them gradually forward. . . . . You must not expect to hear from me before next Saturday, and then only the result of my experiments, as I may not hear the nature of Mr. Stevenson's report for a month after. . . . . I hear various announcements have appeared in the papers concerning our projects. I have had nothing to do with them. The Chairman wished the result of my present experiments not to be made known at present. . . . .

Ever your attached child,

WILLM. F. COOKE.

LETTER XXV.

Tuesday Night, 4th July, 1837.

MY DEAREST MOTHER,

I have not till now had leisure to write a few lines announcing that I have completed my line of wires, extending for 13 miles, and shall have about 2½ miles more, extending to Camden Town and back, laid down as soon as I try my final experiment. I was hard at work all yesterday from a very early hour, and towards evening received a message from several of the London and Liverpool Directors expressing their wish to see any experiments that could be tried along the line, and they would stay in town another day purposely. I promised to do all I could, and worked till 10 at night, and commenced again at 4 this morning. All my wire ends were brought to a table at one end of the room, and neatly arranged over-night; but I would try no experiments till the morning, dreading lest some of my contacts should prove imperfect, and make me fidgetty. Burton Lane was with me by 6 this morning, when I applied my battery, and tried a length of two miles first—all right; then two more with the last—all right; then six miles—all right; then 8, 10, 12, 13, with the same result. All were tried in about one minute, so that the adjusting my instrument and sending 7 messages through a total of 55 miles required scarcely as many seconds. I only arranged the simplest of my instruments, and had all ready by 20 minutes past 9.

I then went home and had a good wash, dressed, with one mouthful of breakfast, and got back by 10 o'clock, the hour appointed. About 20 of the Directors were soon assembled. Mr. Wheatstone could not be present; so I commenced my explanations, and got through them with all the ease and coolness imaginable. I could not have felt less nervous had I been explaining them only to you. I prefaced by saying that I had hastened my preparations not to disappoint those Directors who were leaving town, but did not offer the experiments as a sample of what my Telegraphs were to do, but simply to show that the current of fluid would pass through miles of wire instantaneously, &c. I commenced by putting my Heidelberg instruments in motion, which excited great interest. I then rung a bell, &c., &c., and finally displayed the gradual decrease of galvanic energy in lengthened circuits by transmitting the current first through two miles, and so on to the thirteenth. All expressed themselves satisfied with the principles, and seemed to take the deepest interest in the experiments. They were most courteous in their manner to me, and all came to wish me good-day before leaving. Our final experiments will be made next week, most likely near the end. Mr. Stevenson was present, and played with the instruments more than anyone else. We are to exhibit before him and a Mr. Prevost finally.

We mean to try whether a machine called "Electro-Magnetic" cannot be made to supersede the Galvanic Battery. I have long been anxious to ascertain this point, it being part of my original plan.

I think we shall now have fair play, and if our system is worth anything it will be proved. . . . .

Ever your most affectionate and attached Son,

WILLM. F. COOKE.

. . . . I have this moment received a letter from the Railroad Secretary, requesting me to defer my final experiments till the return of Mr. Stevenson from the country. This will give me time to perfect all our arrangements. The Doctor has kindly given me permission to draw for my immediate demands, which I know you will sanction. In case of success all will be well. I have spent not twenty shillings upon myself during many weeks. . . . .

LETTER XXVI.

16, COMPTON STREET,
BRUNSWICK SQUARE,
9th July, 1837.

MY DEAREST FATHER,

I will begin a line to-day, that I may only have to add a report of our experiments when they do take place. Many thanks for your last letter. I had seen the same advertisement by a Mr. Alexander in the *Times*, shown me by Mr. Creed. There is nothing new in the details mentioned, and they make no mention of alarums, the very pith of our patent. . . . .

We shall determine to-morrow (Monday), what is to be done with the Scotch patent. We have some idea of selling one-third to pay expenses and furnish cash, if anyone will buy, in which case I will buy a share for you, if they sell cheap, equal to the amount of money I must have, which will be the best security I can offer.

The results of our experiments far surpass my expectations. My instruments for bringing a secondary battery into action, at the distance of 14 miles, act under the influence of six plates of my own construction to admiration. I cannot see where we can fail, but dare not be sanguine. Altogether we shall have 16 miles of wire. Professor Wheatstone had calculated upon seven plates per mile, and I upon two. We were both very wide of the mark. I do not think more than one pair per two miles will be required, and those very diminutive. I turned a needle rapidly 90° with no other battery than a tip of wire covered with zinc amalgam, and another tip of copper wire, with a little dilute sulphuric acid between them on my finger, through a circuit of 14 miles. This wonder I have not yet shown to anyone—of course the needle was beautifully suspended.

I reported to Mr. Creed yesterday that we were prepared, and have written to Mr. Stevenson.

Monday Night.

Mr. Stevenson and Mr. Creed have both been with us to-day, and took the deepest interest in our experiments. They wish, however, to see the effect in greater distances still, and I have received orders to get more wire and extend along the road. I had hoped that they might have come to a conclusion at once, but their entering into further expenses is very satisfactory. Mr. Creed expressed himself ready to come down at any moment when we were ready for him. Mr. Stevenson appears to take equal interest in the undertaking.

I need not say that each little delay adds a grey hair to my temples and a wrinkle to my brow, but hope is bright before me.

We have determined that our Scotch patent is safe at present, so take no immediate steps. I will, however, remember your most generous offer, and if I accept it, it will be after mature reflection. . . . .

Ever your most affectionate son,

WILLM. F. COOKE.

. . . . Nothing like a public experiment will take place at present. When the time arrives I will let you know.

LETTER XXVII.

16, COMPTON STREET,
BRUNSWICK SQUARE,
Midnight, 25th July, 1837.

MY DEAREST MOTHER,

A fatiguing day and late hour must be my excuse for rendering this letter as brief as possible, unless I keep it for to-morrow's post, which I may do. . . . . Having a little good news to relate, I am rather anxious, however, to despatch it sooner.

Yesterday Mr. Stevenson witnessed our experiments through 19 miles of wire, extended from Euston Square to Camden Town, and declared himself so satisfied with result that he begged me to lay down my wires permanently between those two points on my best plan, with a view to extending the communication hereafter, if the Directors approved. He wishes us also to have all our instruments on the most approved construction, and I have consequently put several new ones in hand, to be ready, if possible, in a fortnight. He said we must have two or three rehearsals beforehand, that all may go off in "good style." He declared himself a "convert to our system," and seemed quite delighted at the correspondence we carried on at so great a distance from each other, requesting me to send the word "Bravo" along the line more than once. It ended by his desiring me to send an invitation to Mr. Wheatstone to join us, which he politely replied to by saying he would do himself "the honor," &c. Mr. Stevenson was quite childish in the delight he took in the working of our Telegraphs, and seems now to have taken it entirely under his patronage; and a more influential one [*sic*] we could not desire. If he continue to think well of it, we shall have little difficulty to contend against. He has allowed us to go to another £100 expense on account of the Company, and only added as a condition that all should be perfect, for "I do not like "to be laughed at for a failure." He said, "We will have not only "the Directors, but half the nobility and everybody else to see it; "only let all be sure;" and I am sparing no pains to comply with his injunctions. Making out drawings, forming plans, obtaining materials, have occupied my day pretty busily, and I am likely to have a week similarly employed. I have just given orders for 5,000 feet of wood to be sawn in a particular manner, with grooves for the wires, which I am going to have boiled in coal tar previously to laying down. Our wire is all ready.

A variety of notices keep appearing in the papers respecting Electric Telegraphs, but there is nothing in them; some rumours have got abroad respecting ours which have given rise to most of them. The one tried at Munich will not answer on a lengthened line, we having tried similar experiments and proved the insufficiency of the plan. . . . . God bless you, dearest Mother. Good-night.

Your affectionate Son,

William F. Cooke.

Mrs. Cooke has written on the back of this letter, "Glorious news— "delight throughout."

Letter XXVIII.

11, Compton Street,
Brunswick Square,
Saturday Evening [*29th July*, 1837].

My Dearest Mother,

. . . . Mr. Stevenson has left this morning for a fortnight, so nothing can be done till his return. I received such a nice friendly letter from him last night, asking me to show a few experiments to two ladies staying with him, who were shortly going away. I must copy a paragraph:—

"MY DEAR SIR,

"I feel that I am taking a liberty which I ought not to do, in asking "the favor of your repeating an experiment or two to-morrow with Teleg. "apparatus, provided it is not inconvenient to you. I ask this for two ladies "staying at my house, who have their curiosity much excited by the accounts "which I have given them of the result of your former experiments. . . . . Now, "if this proposal be in the least inconvenient to you, I will rely upon your saying so.

"Believe me to be,

"Yours faithfully,

"ROBT. STEVENSON."

Now is not this a nice style of letter to receive from such a man? I was much inconvenienced, but laboured late last night and from four this morning to get all in readiness. Everything acted better than before, and all were delighted. Mr. S. more *heartily friendly* than ever. . . . .

Your attached child,

WM. F. COOKE.

Party after party of ladies came to see us to-day. . . . .

LETTER XXIX.

8, BREED'S PLACE, HASTINGS,
8th September, 1837.

MY DEAR BETSY,

. . . . Monday week I met Mr. Stephenson, immediately after his return from Birmingham, and Wednesday night, at 8 o'clock, was the earliest moment he could fix. I met him in my *den underground* at Camden Town at the time appointed, Mr. Wheatstone, with my brother Tom, being at Euston Square. The Secretary to the railroad (Mr. Creed), and one or two other gentlemen were present with me when our work commenced by the sounding of my alarum, and our enquiring from Tom whether our visitors had arrived, which was politely replied to; and then followed a great variety of communications in the best possible style, the length of wire through which each communication passed being between 14 and 15 miles. No mistakes or blunders of any kind occurred in more than an hour's constant working. Mr. Stephenson continued with me till 10 o'clock at night, deeply interested in the subject; he told me he had thought deeply upon the subject, and was [so] convinced of its practicability, that he would make out his report to the Directors immediately on his return to London from Birmingham, where he was going again the following day; that he was prepared to recommend a much more expensive method of protecting the wires, which would ensure more thoroughly their safety, the cost of which might come to £400 or £500 per mile, or £100,000 to Liverpool; that, though the sum was great, yet the Telegraph was of vital importance to the railroad. Our position, then, is this: Mr. Stephenson not only approves, but will recommend Cooke's and Wheatstone's Telegraph to the Directors for adoption. Our conversation terminated in my mentioning an intention of paying Hastings a visit, and his sending through me a verbal message to his wife, who is residing here. I have consequently become acquainted with her, and my mother and Louisa* have called on her to-day; she and some friends will pay us a visit to-morrow to see some working models that I have brought with me. Mr. S. comes down to-morrow evening, when he is to show me his report, and has offered to make any alterations I may suggest in the plan he recommends to the Directors. All, therefore, is going on as well as can be desired. . . . .

I saw a most incorrect account of our proceedings in the paper the other day, in which Mr. Wheatstone and Mr. Stephenson are represented as trying the experiments, which have been conducted solely under my direction, *I alone* being in communication with the Directors on the subject, and have been the only person responsible from the commencement. I have good reasons for taking no notice of the newspaper reports for the moment. Do not fear that I shall have my full credit at last. I am keeping back, and have my eye on all that passes. . . . .

Believe me,

Yours sincerely,

WILLM. F. COOKE.

---

* The lady referred to as Louisa afterwards became Mrs. Cooke.

LETTER XXX.

16, COMPTON STREET,
BRUNSWICK SQUARE,
25th October, 1837.

MY DEAREST MOTHER,

. . . . I have been vainly hoping that I might be able to give a favourable report of my proceedings, but since my last letter I have absolutely made no advance whatever. The fact is I declined taking any step till my agreement with Mr. W. is finally settled, as all advantages resulting from my exertions render him more difficult to manage. That point settled satisfactorily, I hope to progress rapidly, my friends being all ripe to push the undertaking as soon as I can act without restraint. We had the Belgic Minister and some scientific friends to see some experiments last Monday, which went off with great *éclat*, and afforded perfect satisfaction. His Excellency promised every assistance with his own Government. . . . .

Ever your attached Son,
WILLM. F. COOKE.

Thursday Aftn.

I have nothing more to add about our telegraphic proceedings. After a long conversation with Mr. Wheatstone, we have agreed to leave the case to our arbitrators, who will meet early next week. . . . .

LETTER XXXI.

Friday Night, 3rd Nov. [1837].

MY DEAREST MOTHER,

I have no little pleasure in giving you a more favourable account of my proceedings than the last two letters have contained. After a toilsome and anxious week Mr. Wheatstone has approved of terms favourable to both parties. They still, however, have to be drawn out in legal form, and may yet offer subject of discussion, but his conduct has been so satisfactory during the last week that I have every hope of going on in a most comfortable manner with him. His arbitrator never made his appearance, after waiting more than a fortnight for him, during which time I have intentionally remained quite inactive, as prosperity seemed to increase the difficulties of arriving at a settlement of rights. If the present agreement is carried into effect, I shall be as independent in my proceedings as I can desire; his only control over me in the management of our British patents being in having the power of dissenting to the sale of licenses or the patents in case of not approving of the terms, when, if we cannot agree, reference is to be had to arbitration. I am to have a percentage on all proceeds to cover expenses of conducting the management, then half the remainder, and the sole right to act as engineer. I give up to him several countries on the Continent, into which he has the privilege of introducing the invention for his exclusive benefit, Russia and Austria appertaining to my share, if anything can be done there. This is all as I wished. When the agreement is once signed, I shall push forward again with renewed energy, to make up for lost time. Probably the first step will be to offer the invention to the Gov[t] for their patronage, then secure an Act of Parliament for the formation of a Company. . . . . I am obliged to make advances for the present, or all our works would stand still. I am getting new instruments ready for the inspection of the Company I hope to form. I have not got all the money back I advanced for the L. and B. Railway Co., but that and other sums are perfectly safe. . . . .

Your fondly attached child,

WILLM. F. COOKE.

CLAPHAM, 4 P.M.

Just arrived from Mortlake after a very satisfactory interview with Mr. Hawes. I got down to an eight o'clock breakfast, but was detained there in preparing further documents. I hope early next week our agreement will be signed. . . . .

LETTER XXXII.

9th Nov., 1837.

MY DEAREST MOTHER,

. . . . I went down by appointment to Mortlake last Monday night with Mr. Wheatstone, and discussed with Mr.

Hawes the heads of our agreement. We were occupied at it from 8 till 11 o'clock, when, everything being settled to our mutual satisfaction, our signatures were attached, and the papers placed in lawyer's hands to embody their spirit in legal circumlocution. They contain all I wish, leaving me sole and entire manager in England, Scotland, and Ireland, with this one exception—that, before selling the patents or licenses, I am to obtain Mr. Wheatstone's acquiescence in the price. Nothing, of course, can be more reasonable. I am to have a percentage of 1-10th for my expenses and trouble, dividing the balance equally with Mr. Wheatstone, so that out of every £100 I shall have £55 and Mr. W. £45. I do not take any step till the formal agreement is sealed and signed.

. Mr. Hawes, in whose advice I place the greatest confidence, is anxious that I should immediately establish an office at the West End, as near the Government offices as possible. This of course I cannot do with my own funds, so hope to find others ready to furnish the means. I have a very clever mechanician, who has arranged with 4 or 5 good workmen to commence operations as soon as I give the word. A large manufacturer is preparing machinery to make and furnish the wire, and I have just completed a perfect set of instruments for exhibition before the members of Government, if they are inclined to take it up. These are all the preparations I dare make at present.

Thank Tom for reminding me of the specifications. Mr. W. and I have been constantly at work at them during the last month, and I have to-day completed the last drawing. Tom will be surprised at the labour this part of the subject has required. We have till the 10th December to complete it. Farey, the best consulting engineer in England, is engaged by us, and already has some of the papers. . . . .

I fear we shall hardly have time to go before Parliament this session, as the Government ought to be gained first, if possible. . . . . Depend upon it, I feel pretty confident of success, or would not risk putting you and my dear Father to temporary inconvenience. My brain throbs now and then whilst thinking of what yet remains to be done.

I did not tell you that by our agreement I am to be solely employed in carrying the Telegraph into practice, for which I am to be paid distinctly by the parties engaging me. Mr. W. binds himself not to do so himself, or to allow anyone else to interfere with me. This alone should secure me a handsome income, if the start be once made. In consideration of these privileges I leave to Mr. W. the right of introducing the invention into Belgium, Holland, Prussia, and France, for his exclusive benefit, I having the same privilege in Russia and Austria; so we have divided the spoil pretty freely. Of course we keep all this *private!* . . . .

With affectionate love to the Doctor and old Tom,

Ever your fondly attached child,

WILLM. F. COOKE.

LETTER XXXIII.

13, BEDFORD ROW,

Friday [*Dec.* 2*nd*, 1837].

MY DEAREST MOTHER,

You will think me almost unkind in not writing more frequently, but I never seem to have a moment to myself. Drawings, specifications, letter-writing, and running all over London to attend the pleasure of different persons, and visits at Bedford Row for my own, occupy me thoroughly. Dear Louisa has sketched you out the result of yesterday's work, which was very successful. Alexander behaved very handsomely, and did not press his opposition, so our Scotch patent proceeded last night, and will be sealed in about a fortnight, if all go well.

I have arrived at nothing definite with Brunel, but a prospect seems to be opening on his line (the Great Western). I am too experienced however in disappointments to be very sanguine. My position with Russia seems promising, and if I can succeed in interesting Baron Linglitz's pocket it may turn out a lucrative affair.

Poor Wilson! How he would have rejoiced in the promising

position of our affairs! His friends and family seem as warmly interested as ever, especially his partner, Mr. Lancaster.

Our English specification will be ready on Monday, but is not due till the 10th, so I shall have time to think it over and prepare the American one from it. Those affairs settled and the Scotch patent out, I shall be at leisure to push the undertaking forward. . . . .

Ever your fondly attached Son,
WILLM. F. COOKE.

LETTER XXXIV.

BEDFORD ROW,
Tuesday morning [*20th Dec.*, 1837].

MY DEAREST MOTHER,

We are off this evening for Treeton, and never did schoolboy anticipate a holiday with greater delight. I must be back again with the beginning of the year, however, and hope 1838 may open brilliantly for me.

Yesterday I despatched my papers, drawings, and models for America, and thereby relieved my mind of a considerable load. The Scotch patent was granted on the 12th, six months to a day after the English one. . . . . The L. and B. Railway C[y] have offered me the instruments [used at the experiments] at half-price, and declared, when twelve in council, the most friendly feeling both towards me and the undertaking. . . . .

I suppose you will all quarrel with me when I tell you that I have been exerting myself to prevent the publication of our specification till February next. I feel more than ever convinced that silence has been sound policy, and mean to persevere in it yet for a time. I have many active influential friends working in good earnest, all of whom seem to feel confidence in our undertaking, from the steady unpretending manner in which we have proceeded. . . . .

Ever your affectionate Son,
WILLM. F. COOKE.

Letter XXXV.

[12*th Jan.*, 1838.]

My Dearest Mother,

. . . . I am an idle man at present, in daily hope (which hope is nearly a month old) of finding myself but too busy. In hourly expectation of hearing from Mr. Brunel, I am afraid of taking any other step. In the meantime, I have everything in preparation for the important step of forming a Company. . . .

Did I tell you that Mr. Brunel took me in his carriage to Maidstone on Thursday week, to view his railroad and works? We had a most delightful trip, and did not return to town till past midnight. . . .

Ever your affectionate Son,

Willm. F. Cooke.

Letter XXXVI.

16, Compton Street,
Brunswick Square,
10*th March*, 1838.

My Dearest Mother,

. . . . I have been in daily expectation since my last, of entering into final engagements with the Great Western Company, and all went on favourably till Tuesday, when the sub-committee appointed on purpose made such unreasonable alterations in the heads of agreement, which we had previously been over two or three times, that I determined to withdraw altogether, and wrote accordingly to that effect. I suspect they did not anticipate such independence on my part, or they would not have driven me to it. We parted, I hope, perfectly good friends. At my next visit I will bring with me all the papers connected with this transaction, and I think you will all agree that I acted only with becoming spirit. I am now again without any prospect, and having delayed my experiments at St. Katherine's Docks on their account, I now proceed with them. The suspense in which I have been kept by the Great Western Company rendered it impossible for me to write. Do not mention this second discomfiture, as, if known widely, it may prove very prejudicial, even though I was the party to withdraw, which, of course, no one will believe. I have this satisfaction, that the Chairman, Mr. Sims, told [*sic*] before the whole committee that they must have a Telegraph, and that they believed ours to be the best—an opinion strongly supported by the Engineer. . . . .

Ever your attached child,

WM. F. COOKE.

LETTER XXXVII.

13th March [1838].

MY DEAREST MOTHER,

This must be a short, though I am sure it will prove a welcome, line, as the bearer of better news than I had hoped to give. The Great Western Company have renewed the correspondence under the most flattering circumstances to myself, and the most propitious to the invention. How it may terminate I cannot pretend to say. The step that has led to this change of position was a neck-or-nothing one. The returning their amended proposals without a remark astonished them all. The manner in which I did it, however, has so far from displeased that they have received me again most cordially.

I had a long interview (from 10 p.m. till past midnight) on Monday with Brunel, who had that day returned from Wales. He let me have a peep behind the scenes, and it appears that the Company had never before been treated so coolly, and were absolutely pleased with the novelty of the occurrence. Of course I had taken care not to offend their pride. I was to have seen the Secretary yesterday, but he was from town, and have only now returned from a two hours' most satisfactory interview. The agreement is again in my hands for alteration. If we come to terms, they will now be satisfactory, though perhaps not very lucrative, yet still handsome; but once started under Brunel I may make money elsewhere. I think I have secured my being treated as a gentleman by everyone connected with the Company. . . . .

Pray keep all that I have written to yourselves, as I would not for the world that anyone should know that I am so gratified with what has occurred, or had so fully calculated upon the result of a bold step. . . . .

Your affectionate child,

WM. F. COOKE.

LETTER XXXVIII.

3rd April, '38.

MY DEAREST MOTHER,

A few lines of good news must compensate for a hurried and brief letter.

I have this day concluded my agreement with the Great Western Railway Co. The terms are quite prospective—no cash before the 1st Jan., 1839—but, what is most important, they try the Telegraph immediately. The first trial is to Drayton from Paddington, 13 miles, which, if approved, is to be extended to Maidenhead immediately. On the 1st Jan. they determine whether they will take a license, and to what extent. They have the refusal of a considerable district, and if they take all they now most eagerly bargain for I shall at once be an independent man. I cannot in a letter explain even the general nature of the terms, but you shall see all when we meet.

I am going down to the North on Monday, paying Liverpool, Birmingham, Manchester, and most probably Hull and Newcastle, a visit, with 4 days' rest at Treeton. I had a large party of first-rate City men in the Baltic to-day—they seemed deeply interested. I think things are really starting now.

The Great Western pay me a certain sum per mile for laying it down. How much I gain by it will depend upon my own good management—I hope £200 or £300 at least. I commence as soon as I can obtain the materials. Brunel nearly met with a fatal accident the other day, but I rejoice to say the report this even^{g} is favorable. . . . .

Our Scotch specification goes in on Friday. It has cost more than £120, besides the cost of the patent. Stamps alone are £38. . . . .

I shall most likely change my quarters, and get rooms where I can show the apparatus without trouble. These experiments have cost me most dear, in money, time, and trouble, but they have not been in vain. . . . .

Ever your affectionate Son,

WM. F. COOKE.

LETTER XXXIX.

TREETON*

[18*th April*, 1838].

MY DEAR FATHER,

. . . . I have need at the present moment of some ready cash, and shall be much obliged to you for the re-loan of £100 for a *month or six weeks.* I am unwilling to draw on the Company till I have something to show, otherwise they will honor my drafts as I make them. You will think I have taken a bold step, but it has been on very safe calculations. I hope to be at work the first week in May. If convenient, will you remit the cash to Burton Lane, who will have my directions for applying a portion of it.

The Scotch patent, which was enrolled on the 10th inst., cost me more than £80 beyond the stamp costs, no mention of which had been previously made to us, and till I can see Mr. Wheatstone I am obliged to furnish all the money. Our patents have already cost upwards of £800.

Davy has again been in the field against us, and we have given him two signal defeats—the last, I think, a final one. . . . .

Legal difficulties respecting the nature of license to be granted to the Great Western Co. occupied me during the week previous to leaving town. The Company throughout our correspondence, since the misunderstanding was cleared up, have shown the most friendly and cordial feeling towards me. Brunel, I am happy to say, is rapidly recovering from his accident. . . . .

Ever your affectionate Son,

WILLM. F. COOKE.

---

* Treeton was the residence of his intended wife

LETTER XL.

TREETON
[18*th April*, 1838].

MY DEAREST MOTHER,

. . . . I have come down fully prepared for a tour, if circumstances render it desirable; but I am inclined to think that I shall defer my visit to Hull, Newcastle, and Liverpool, till after the Telegraph is carried out to Drayton. . . . . My preparations for that distance are going on rapidly. . . . .

I could not help smiling at your observation about my travelling last Tuesday, when *you* had spent the day in making *fifty* calls, dear old Mother!

I have been taking a bold step in contracting to lay down the Telegraphic wires to Drayton for £2,145, or £165 per mile. I tried to avoid it, but was so hard pressed by the Company that I acceded, warning them repeatedly that they could do it themselves for very far less. I cannot fail to make a good thing of it, having contracted with very responsible parties to supply me with my covered wire, tubing, &c., at much less than the estimated cost. I am also promised every possible assistance from the resident engineer, Mr. Clark, a great friend of mine. All this *entre nous*. If the Telegraph be carried on to Maidenhead, I have the option of contracting for the same on the present terms, which are 50 per cent. above the estimate. . . . .

I saw Mr. Robert Stephenson the Saturday before leaving town, and rather hoped to see him and his father here; their visit would decide me on the advisability of proceeding further north at the present juncture. . . . .

Ever your affectionate Son,
WM. F. COOKE.

LETTER XLI.

19*th May*, '38.

MY DEAREST MOTHER,

. . . . I have been waiting, not in daily, but hourly, expectation of having some certain news of my proceedings with the Great Western Co. to relate, but day after day has passed heavily away, till my patience is nearly exhausted. I had an interview with Mr. Saunders and Mr. Brunel on Thursday, and again on Friday, when I gave in the agreement, duly drawn up, and altered, as I hope, to their minds. On Tuesday I was to have had an answer, but none has yet reached me. What they are doing with it, unless forgetting its very existence, I cannot surmise. Their road is to be open on the 5th of June, and it is probable they have little time to think of anything else.

In the meantime I have had nothing to do, so went in search of lodgings near the Great Western Terminus at Paddington, in case of my needing them, and have seen some very much to my liking; but every positive step must be suspended till I hear more. Suspense is ill to bear during an idle time. . . . .

Ever your affectionate Son,
WM. F. COOKE.

LETTER XLII.

16, COMPTON STREET,
BRUNSWICK SQUARE,
28th May, '38.

MY DEAREST MOTHER,

. . . . I commenced business early this morning, having received the agreement from the Directors on Friday night. Very busy about it all Saturday, but did not venture to write, as I knew not what the results might be. I was with Mr. Brunel this morning at 9; received directions to commence forthwith from him, if I chose to approve of the agreement, or whenever, in fact, I saw right to sign it. I had already made up my mind. By 10 I met Wilson, who had the agreement regularly drawn up and stamped, which we finally settled. Called on Mr. Wheatstone by 11, talked it over, signed it, and Wilson witnessed it. At 12 precisely met Mr. Saunders and the Directors, who ratified it, so now this mighty affair is settled. Now comes the proof of the patent.

On returning home I found your letter, and am now answering it before I ride to Paddington to select my temporary office, which Mr. Brunel recommends me to establish in one of the arches of their principal bridge, and in all respects likely to suit me, being both roomy and light. Orders are already going round to all my workmen to re-commence operations most accurately, and I think it possible three miles may be done with fine weather by the end of next week, but I may have to wait for timber, as Mr. Brunel seems inclined to change his plans. This I shall know on showing my plans to-morrow morning. . . . .

Ever your affectionate Son,

WM. F. COOKE.

LETTER XLIII.

31*st May*, '38.

MY DEAREST MOTHER,

I find it would be impossible for me to leave town to-night, without interfering seriously with my proceedings. I get into a new office they have built me to-day, and I should not be out of the way to-morrow, either for appearance sake, or for the advancement of my preparations, Saturday being the day for engaging men upon the works.

I was invited yesterday to join the first experimental trip on the Great Western Railway, which I could not but accept, and thereby lost a day; but I spent it most delightfully—no accident or inconvenience of any kind, and a charming day. The party consisted of 250, so I was really complimented by being remembered. Many of the first men of the country were there, in the engineering, architectural departments, &c., &c. . . . .

I look upon this drilling as a jewel of a plan—1st, because this will in consequence be no experimental trying of plans on the permanent line; 2nd, no interference from lookers-on, and no awkwardness or misunderstanding when under the eye of the engineers and other men. Besides, the plan is a novel one which Branel has decided upon, and I have thought less upon it than on any other. I mean to lay down about a quarter of a mile in the garden over and over again, till each man knows his duty. The materials will do again on the main line, except the wires, which will have their jackets worn out; and even they can be burnt, cleaned, and re-covered. It will only cost me £4 or £5, and the expense to Drayton is £3,159. Something of an undertaking, you see! but I shall be too late for the coach.

God bless you all.

Ever your affectionate Son,

WM. F. COOKE.

LETTER XLIV.

28, CRAVEN PLACE, BAYSWATER,

8th June, '38.

MY DEAREST MOTHER,

Whilst I have the power to write a line I will, and add more if possible afterwards. Your letter reached me on Tuesday night, and I confess the next morning I had forgotten that any part required immediate attention. You will not doubt my being busy every hour when I tell you a letter from dear Loui requesting an answer by return of post has been by me two days—but more of this presently. . . . .

I rode nearly 30 miles yesterday, giving final orders for the delivery of goods, and making of small pieces of machinery, &c., &c., all of which required thought and calculation. Our plans are entirely changed; and I am to lay down the wires in iron pipes. In consequence, all the instruments and tools before made are useless; and one fresh set is nearly complete. The plan now adopted is the most perfect I can desire, but the cost I always feared would be the objection. I have between £600 and £700 worth of goods coming in to-day and to-morrow, and 15 or 20 mechanics at work, constantly requiring direction, so that I have not one moment's rest; and am drawing and making calculations half the night. . . . .

I have just tried a fresh series of experiments to the extent of nearly 40 miles, and with more satisfactory results than before. Therefore to anticipate a failure would be absurd. The Company have embarked £3,000 in the extent of line to Drayton alone, and then will follow it up by £2,000 more to Maidenhead within a fortnight after getting to Drayton; and they have engaged my services for one year afterwards, Mr. Brunel and Mr. Wilson determining the amount of salary.

With these prospects, would you not justify my taking the step I have proposed, should the emergency demand it? You will believe my mind has little rest; but, with the exception of my anxieties respecting Treeton, all is delightful and going on perfectly to my satisfaction, and the worry of so many things to think of will be over in a few days—as soon, in fact, as we start. I must now to work. The hurry in which I have written, and interruptions, have prevented my being very concise or clear, but you will be able to see what is passing in my mind; and to you and my dear Father I have not, I know, a difficult case to make out.

I have got charming lodgings, close by Kensington Gardens, similarly arranged, but much better and handsomer than in Compton Street, and one shilling a week cheaper. Green fields, gardens, and country around, and close by my work. I am in famous health.

This letter is not signed. Mrs. Cooke has written over the address:

"God be with my blessed boy, and prosper this mighty discovery. "Poor Louisa! How their minds must be tried! May it please God to spare "her dear mother to witness her happiness or it will be sad!"

Letter XLV.

Sussex Cottage, Slough,
28th November, '39.

My Dearest Sister,

. . . . I have greater pleasure in writing to you now, as I can announce the improved prospects of the Telegraph, Mr. Robert Stephenson having engaged me to lay it down on the

London and Blackwall Railway. The terms, which have been approved and confirmed by the Directors, are very liberal. There are many gratifying circumstances connected with this engagement. One is, my having been sought by my *first Patron,* after losing sight of him for nearly two years. Another, that my plan of working the Telegraph on that line (which is quite novel) so completely meets Mr. Stephenson's views that he will be enabled to extend the accommodation of that railway to several localities of the densely peopled district through which it passes. I hope to finish this short line in March next.

The Telegraph on the Great Western has given great confidence and satisfaction. It has been working for two months almost incessantly, from 8 in the morning till 9 at night; the signals being recorded by a clerk of the Company, under their Secretary's directions.

George will be sending a parcel to Burton Lane in a few days. If you can find some drawings of my original instrument, and a book of description which Charles took abroad with him, I should be glad to receive them at the present moment included in the same packet. . . . .

Ever your attached Brother,

WILLM. F. COOKE.

I had hoped to have
had my Experiments made public to day, but
have not till the Patent is out – as one days impatience may ruin all. The King's health is still
so precarious that he can transact no business.
A report was very prevalent yesterday that
he was dead, but contradicted in the Eng–
Again adieu with my affte wish for many
and many a happy return of this day, believe me
my own dear Mother your attached and grateful son
William F. Cooke

Abroad for the 18th of
June

The material originally positioned here is too large for reproduction in this reissue. A PDF can be downloaded from the web address given on page iv of this book, by clicking on 'Resources Available'.

Cooke's & Wheatstone's

Patent Signed by His Majesty, and receiving the Great Seal this day June 10th!!! 1837 for Electric Telegraphs Alarums

I had intended to send you some small ~~parcel~~ present, but now send this instead

The material originally positioned here is too large for reproduction in this reissue. A PDF can be downloaded from the web address given on page iv of this book, by clicking on 'Resources Available'.

My latest wish is, that my funeral should be most modest and inexpensive, with a simple stone at "my head" inscribed simply thus —
"Willm Fothergill Cooke.
died. [ ] 187– "
Nothing more — whatever —
And if my name should be recorded on our family vault at my Oatlands Churchyard, and let it be — simply —
name / W. F. C. — in Memoriam —
Born —— May 4th 1806
died and buried in Churchyard at Farnham
—— 187–
May 4th / 1879 —
~~[illegible] 1878~~
Willm Fothergill Cooke

The material originally positioned here is too large for reproduction in this reissue. A PDF can be downloaded from the web address given on page iv of this book, by clicking on 'Resources Available'.

Memoir of
Sir WILLIAM FOTHERGILL COOKE.

By LATIMER CLARK, F.R.S., M. Inst. C.E.,
*Past-President, Institution of Electrical Engineers.**

Sir William Fothergill Cooke, an Honorary Member of this Society, who died on the 25th June, was one of those men who, though like Ronalds, little known or regarded by his contemporaries, has contributed an important chapter to the history of the industrial and scientific progress of the nineteenth century, and has added largely to the fame and reputation of his country. He has at the same time made a name for himself which will grow greater as we recede from it, and which is probably destined to endure through distant ages.

As Stephenson, though not the actual originator of the Railway System, has inseparably associated his name with its introduction, so Cooke, though he contributed but little to the invention of telegraphy, has earned for himself the title to claim the honour of being "entitled to stand alone as the gentleman to whom this country is indebted for having practically introduced and carried out the Electric Telegraph as a useful undertaking"—an honour which was awarded to him after an almost judicial investigation of the facts by Sir Marc I. Brunel and Professor Daniell, at the time of the arbitration of the differences between himself and Professor Wheatstone in 1841. This, too, was six years before he crowned his labours by establishing the Electric Telegraph Company, the earliest telegraphic association in the world, and which, under his direction and that of his friends, soon grew to be the largest and most successful.

He survived his contemporaries Ronalds, Wheatstone, and Morse by only a few years, and with him has passed away the last of that little band of men who have been among the foremost in conferring upon the world the practical benefits of the Electric Telegraph. More fortunate than many benefactors of mankind, they had the satisfaction

---

* *Journal of the Society of Telegraph Engineers*, vol. viii, p. 361, 1879.

of living to see their labours completed, and the wildest dreams of their early anticipations realised more than tenfold. Even to those who well remember the pre-telegraphic era, when it took days and weeks to communicate with distant countries instead of minutes, it is difficult to realise the extent of the change which has been produced in human existence, by the joint operation of the railway and the telegraph; to those who come after us, it must be impossible. Others before him had striven earnestly to attract the attention of their countrymen to the importance of the discovery which lay neglected at their feet, but gave up the attempt in despair, or were repelled like Ronalds, and driven to retire from the subject in disgust. Cooke was more fortunate, not only in possessing a greater amount of energy and determination, but in falling upon days of great commercial activity, when capital was abundant, and when railways were beginning to spread rapidly, and were enabled to lend a helping hand to the young and rising invention, from which they were soon destined to receive such ample benefits in return. He was further fortunate in making the acquaintance and obtaining the co-operation of many of the most influential and eminent scientific men of his day.

Sir William Cooke was born near Ealing, on the 4th May, 1806. His father, Dr. William Cooke, who died in March, 1857, was a doctor of medicine and professor of anatomy at Durham, and was appointed Reader in Medicine to the newly organised university there, where he began his lectures in 1833.

His son was educated at a school at Durham, and was subsequently sent to the university at Edinburgh. In 1826 he was appointed to the East Indian Army, in which he held a variety of staff appointments. After nearly six years' service, and while in the Madras Native Infantry, he returned to England on furlough, on account of the state of his health, and he soon afterwards relinquished his appointment. In 1833 and 1834 he betook himself to Paris and attended lectures there, in order to devote himself to those more scientific and practical pursuits for which he felt himself fitted. He studied anatomy and physiology, and practised with great ardour the pursuit of modelling anatomical dissections in coloured wax—an art in which he eventually acquired great skill. In the spring of 1834 he returned to Durham, and prepared a series of models with which his father used to illustrate his lectures at the Durham University; he had even formed the intention of founding an anatomical museum there. In the summer of 1835 he accompanied his parents on a tour to Switzerland. Ascending the Rhine, they visited Heidelberg. Here Professor Tiedeman, the director of the then existing Anatomical Institute, offered to assist him in procuring the means to make his wax preparations. Accordingly, Mr. Cooke returned in the month of November from Berne to Heidelberg, where he took lodgings in the Stöckstrasse, in the house No. 97, at that time belonging to the brewer Wilhelm Speyrer, but now to the brewer Georg Muller.* It bears the strange inscription, "Bierbraueri zum neuen Essighaus." There had been before vinegar works in that house. As Mr. Cooke was not permitted to make here, in the cleanly kept apartments, anatomical dissections, he hired a room in the same street, nearly opposite, in the house of the gardener Schwarz, No. 58, now belonging to his grandson, Ferdinand Koch. Here he was during the winter so active, that he was able to despatch four cases full of models to his father at Durham.

"In the present Anatomical Museum at Heidelberg, I have," says Dr. Hamel, "under Nos. 382, 383, and 628, found three wax models made by Mr. Cooke during the winter mentioned; the last is marked W. F. C. Dunelm."

In the beginning of March, 1836, Mr. Cooke heard accidentally from Mr. J. W. Rizzo Hoppner, who afterwards became a member of Mr. Robert Stephenson's engineering staff, that the Professor of Natural Philosophy there had an apparatus with which he could signal from one room to another—this was Baron Schilling's telegraph.† Mr. Hoppner's father was an intimate friend of Lord Byron, and in one of his letters to his publisher, dated February 20, 1818, he remarks: "Mr. Hoppner has been made the father of a very fine boy; mother and child are doing very well indeed." And the poet wrote four lines upon his birth, which have been metrically translated in ten other languages.‡ The Professor was Geheime Hofrath Muncke. He had, in the upper storey of the former convent of Dominicans where he gave his lectures, and where also he lived, suspended wires for telegraphing out of the cabinet into the auditory. "I have examined these localities," continues Dr. Hamel, "and the rooms are now quite empty." From the year 1850 to 1852 the house had served as a

---

* See Dr. Hamel's "Historical Account of the Introduction of the Galvanic Telegraph." London, August, 1859. A most interesting little work.

† Baron Schilling died at St. Petersburg, 7th August, 1837.

‡ Murray's "Poetical Works of Lord Byron," p. 571.

military barrack. Mr. Hoppner took Mr. Cooke on the 6th March, 1836, to Professor Muncke's lecture room.

Mr. Cooke himself thus writes, in 1840, of this period:* "About the 6th March, 1836, a circumstance occurred which gave an entirely new bent to my thoughts. Having witnessed an electro-telegraphic experiment exhibited about that day by Professor Muncke, of Heidelberg —who had, I believe, taken his idea from Gauss—I was so much struck with the wonderful power of electricity, and so strongly impressed with its applicability to the practical transmission of telegraphic intelligence, that from that very day I entirely abandoned my former pursuits, and devoted myself thenceforth with equal ardour to the practical realisation of the electric telegraph. Professor Muncke's experiment was at that time the only one I had seen or heard of; it showed that electric currents being conveyed by wires to a distance could be then caused to deflect magnetic needles and thereby to give signals. It was a hint at the application of electricity to telegraphic purposes, but provided no means of applying that power to practical uses. His apparatus consisted of two instruments for giving signals by a single needle, placed in different rooms, with a battery belonging to each; the signals given were a cross and a straight line marked on the opposite sides of a disc of card fixed on a straw, at the end of which was a magnetic needle suspended horizontally in galvanometer coils by a silk thread." Either the line or the cross could be exhibited by its motion according to the direction of the current. The apparatus was worked by moving the ends of the wire backwards and forwards between the battery and the coils. "Within three weeks after the day on which I saw the experiment, I had made, partly at Heidelberg and partly at Frankfort, my first electric telegraph of the galvanometer form, which is now at Berne." This telegraph had three needles and six wires in three distinct circuits, with three keys and a rudimentary switch; it gave 26 signals.

Mr. Cooke thus writes to his mother on this subject in a letter from Heidelberg, dated April 5th, 1836, to Mrs. Cooke at Berne:—

"You must know that for some weeks past I have been deeply engaged in the construction of an instrument which I believe may prove of sufficient importance, should I succeed in bringing it to practical perfection, to merit a visit to London. Determined to satisfy myself

---

* "The Electric Telegraph: Was it Invented by Professor Wheatstone?" By William Fothergill Cooke. London: Smith & Son. 1856. 2 vols., 8vo. Vol. ii., p. 14.

on the working of the machinery before I went any further, I prepared to make a model, and, being unable to obtain the requisites at Heidelberg, I sought them at Frankfort. Whilst completing the model of my original plan, others on entirely fresh systems suggested themselves, and I have at length succeeded in combining the *utile* of each; but the mechanism requires a more delicate hand than mine to execute, or, rather, instruments which I do not possess. These I can readily have made for me in London, and by the aid of a lathe I shall be able to adapt the several parts, which I shall have made by different mechanicians for secrecy sake. Should I succeed, it may be the means of putting some hundred pounds in my pocket. As it is a subject on which I was profoundly ignorant till my attention was casually attracted to it the other day, I do not know what others may have done in the same way; this can best be learned in London. You see I am very mysterious at present, and think it very prudent to continue so; nevertheless, to you, dearest mother, if it were your wish, my plan and instrument should be explained now, though I think without better drawings than I could make you would scarcely comprehend me. As I do not wish my motives for re-visiting London to be generally known, you had better, in mentioning it to any friends at Berne, state that private business requires my presence, and allow them to ascribe to modelling or what they please the sudden change of my plans."

The original plan he here refers to was the galvanometer telegraph, but the "fresh systems" which have suggested themselves, and which are fully described later on, refer to his chronometer escapement or "mechanical" telegraph, which consisted of two musical boxes with lettered dials working synchronously like the Ronalds telegraph; the mechanism being started and stopped by an electro-magnet and detent opposite the required letter. This instrument, which is frequently referred to hereafter, and which is described and figured in the two volumes before referred to,* occupied his thoughts from this time till May, 1837; but, owing to his want of the requisite mechanical skill and assistance, it never at any time performed satisfactorily. In the hands of others, however, the chronometric principle has been most successfully extended and brought into practical use, as in the Hughes instrument.

Mr. Cooke says:† "Before the end of March, 1836, I had invented the alarum, worked by clockwork mechanism, by the removal of a

* Cooke, vol. i., p. 31; vol. ii., 217.

† Cooke, vol. ii., 17.

detent with a magnet in close proximity to an armature of soft iron forming the tail end of a lever detent. When an electric current passed round the voltaic magnet the magnetism which was for the moment excited in it attracted the tail end of the lever, and by so doing drew its detent out of the clock. It was replaced by a reacting spring or balance weight."

He adds: "The first idea of it suggested itself to my mind on the 17th March, 1836, during my journey from Heidelberg to Frankfort, when reading Mrs. Somerville's work on the 'Physical Sciences.'"*

On the 14th of April, he writes to his mother: "On Monday my packages started for Berne by the waggon. They consist of a packing case containing models, and a small box containing instruments, &c.;" and on the 15th his passport shows that he was himself on his way to London, where he arrived on the 22nd April, and at once commenced his operations.† On the 26th he writes:—

"I have at length found myself once in London again. I dined with Mr. Fergus, M.P., and there met Mr. Hoppner, who gave me your last letter, which had reached Heidelberg after my departure. Relative to my being in England, and the cause, I will explain the whole plan to D. when we are together, but object to so doing on paper, or its being generally known to our friends, as in case of failure (always a strong probability), remarks, warnings, and advice are the more overwhelming. I have written to Tom begging him to prepare himself with one branch of the subject; and when I have finished my instruments, I will divulge my whole plan to him, and perfect together such papers and statements, &c., as will be necessary ere I proceed further. I have not yet fixed upon any patron (a very important consideration), as the commercial and political world are equally concerned. I have a choice between Government and the mercantile potentates. Whatever course I eventually take will be directed by the best advice," &c.

On the 2nd June he says: "I heard from the Doctor a few days ago. He wrote me a most delightful and affectionate letter. I have explained the instrument and its uses to Robert, who will explain all to you."

---

* "The Connection of the Physical Sciences," by Mrs. Somerville. London, 1834. There does not appear to be anything in this work which could have directly suggested the machine devised by Mr. Cooke.

† Cooke, vol. ii., 18.

And on the 6th: "Tom* came up to dinner on Saturday, well pleased to shake off for a time the monotony of Cambridge. We commenced work immediately, and by mid-day yesterday he became fully master of the affair. Although it be impossible to give you even an imperfect idea of it in writing, he will give you his idea of its importance and practicability. I am anxious to complete my instrument as quickly as possible. I fear two months will pass away ere I can hope to lay plans and instruments before the public. . . . . Tom and I are going to the Adelaide Gallery to study various scientific instruments connected more or less with our object in hand. . . . . It is impossible to describe the comfort and delight of having him to consult and talk to. My confidence has risen sevenfold."

On the 21st July, 1836, he first mentions his sketch or prospectus of his invention, which was intended to be published as a pamphlet, but was not printed. It is headed "Plans for establishing a rapid telegraphic communication for political, commercial, and private purposes, in connection with the extended lines of railroads now in progress between the principal cities of the United Kingdom, through the means of Electro-Magnetism. By W. F. C." It is given at length in vol. ii., at p. 241, and is an interesting document from the period at which it was written. His instrument is to give 60 letters, numerals and signs, and he describes the use of his testing galvanometer or "detecter" much as Ronalds had done in 1823. He speaks of cipher codes, and points out the benefits such a system would confer on the Government, the seaports, the railways, the Stock Exchange, and the Post Office. The letter is as follows:—"July 21.—I think that three months must elapse ere I can know the fate of my projects; one instrument will be shortly so far advanced as to enable me to see whether it answers my expectations. I must then have a second made, and both finished before laying them before the public. The difficulties you allude to of securing the wires were the first I surmounted before thinking further of the instrument, and having succeeded to my content in that respect, I then worked out the remainder. My prospectus is ready, but I am about to send it down to Mr. Chevalier, with detailed drawings, for his judgment and correction. . . . . I will shortly send you my prospectus, but its length will occupy two or three very closely written sheets, and I have not time for that at present. I will only

* His brother, the Rev. Thomas Fothergill Cooke, M.A.

say that if the wires were broken anywhere between London and Portsmouth, I would find the injury out and repair it in less than eight hours. When you see the prospectus and explanations attached you will be convinced of that. A guard or watch is out of the question, but the mode of discovering the injury when done is both rapid, easy, and decisive. Still, I beg you seriously not to be too sanguine of my success. I do not know yet that my instrument will answer, and then very probably it may never be used during my life. I fully believe that the day will come when such a means of conveying intelligence will be employed."

Mr. Cooke states, at vol. ii., page 22, that he had at this period given considerable attention to the escapement principle, or step-by-step movement, which was subsequently brought to such perfection by Wheatstone, and he gives, at plates 2 and 7, a drawing of his instrument.

By August, 1836, Mr. Cooke's account books show that he had already expended £361 14s. 8d. on his experiments.

On September 2nd he says: "My model in progressing. I hope to see it work before I cross for the Continent; it is now so near completion that I may speak with confidence of its answering its destined purpose. It is a very showy-looking affair; the Doctor was much pleased with it."

On the 7th October, writing to his mother at 7, York Buildings, Hastings, he says: "My clockmaker has again disappointed me. I called in Clerkenwell on Monday evening in full expectation of finding that everything had been completed several days; but he told me the balance work had been broken by the running down of the works, so I must wait again." And on the 22nd October: "My instrument was to have been finished this morning, but upon calling I found that a wheel in the escapement movement was wrong, and had to be altered. I am to have it on Monday. You would be astonished at the stoicism with which I bear these repeated and endless disappointments. The instrument looks beautiful—I hope it will go as well."

On the 24th November he gives Mrs. Cooke a very interesting account of his first visit to Faraday, to whom he explained his plans and his difficulties: "My motive for writing to-day, at a moment when much pressed for time, is to give you a little more cheering news about my instrument than I could do in my last. At that time I had quite despaired of success, and only wished for a justifiable opportunity of giving up, from an impression that I had detected an error in the principle itself. I could not at the moment make up my mind to tell you. Tom will understand and explain this. I had good reason for believing that it was a well-ascertained fact that the galvanic fluid only imparted a magnetic quality to cold iron or an electro-magnet when its course was short, and that the shorter the course the greater the attractive power, hence magnets were made with several short thick wires for experiments requiring *quantity*, and one long thin wire when intensity was required;* to set this point at rest I got an introduction (through Dr. Uwins) to Faraday, the 'king of electro-magneticians.' He received me yesterday, and I asked him to give me his opinion upon the instrument, and in the kindest manner he proposed calling this morning for half an hour, which he did, but stayed an hour and a quarter, entering with great interest into all the details. He finally gave me as his opinion that *the principle was perfectly correct*, and seemed to think the instrument capable when well finished of answering the intended purpose. He would not give me an opinion as to the distance which the fluid might be passed in sufficient quantity, but observed that if it were only for 10 or 20 miles it can be passed on again. He said in reply to my question, 'I am afraid of inducing you by my advice to expend any large sum in experimenting, but it would be well worth working out, and a beautiful thing to carry on in this manner a conversation from distant points; and the instrument appears perfectly adapted to its intended uses.' Now I consider this highly satisfactory. He took his leave in a most friendly manner, but in a way which induces me to think that he does not mean to take any further step in the affair. I asked his advice as to my way of proceeding in bringing it before the public when completed, but he declared his inability to advise me. . . . . The difficulty of arranging the escapement I shall be able to overcome without a doubt."

On the 28th November he writes: "I have given Moore my new

---

* See Peter Barlow, F.R.S., "On the Laws of Magnetic Action as Depending on the Length and Dimensions of the Conducting Wire."—*Edinburgh Phil. Journal*, January, 1825. Professor Barlow says: "In a very early stage of electro-magnetic experiments it had been suggested that an instantaneous telegraph might be established by means of conducting wires and compasses. . . . . I was therefore induced to make the trial, but I found such a sensible diminution with only 200 feet of wire as at once to convince me of the impracticability of the scheme. It led me, however, to an inquiry as to the cause of this diminution, and the laws by which it is governed." Mr. Cooke had perhaps seen this paragraph.

plan for the escapement. He acknowledges its superiority to our former plans. Faraday could not speak as to the distance the fluid can be conveyed, as he does not know himself; experience alone can prove this. The most minute quantities have been conveyed 4½ miles without any perceptible abatement of their force. The Germans are of opinion that a thousand miles would make no difference. Faraday says that remains to be proved, but does not deny the position. I am still of opinion that the only way to ensure success is to part with some of the advantages to be derived from it to those who are to bring it forward. You are right, dearest mother, not to be too sanguine; the projectors of an original undertaking hardly ever yet reaped the benefit of it. The steam engine, gas, &c., ruined the inventors; the electro-magnetic telegraph shall not ruin me, but will hardly make my fortune."

On the 8th December he writes to his father, Professor Cooke, M.D., Hastings, and for the first time mentions Mr. Walker, who afterwards aided him so effectually by introducing him to railway directors and others. These letters serve to show the business-like manner in which he conducted the affair. "You will be surprised," he says, "to see my handwriting again so soon, but an idea has crossed my mind, and, as you can perhaps advise, I will not lose a post in imparting it to you. I have heard nothing more from Faraday, and in this dilemma I thought of Mr. Walker, whose tastes, connections, situation in the neighbourhood of railroad, and influence, and friendship for you render it possible that he may interest himself about my telegraph. . . . . If you think Mr. Walker a likely person, you may state the circumstances connected with my plans. My invention has a national and commercial object in view. The instrument and principle, coming peculiarly under Faraday's department, have been approved by him. I am perfectly willing to explain all my views, and obtain, if possible, a written opinion from Dr. Faraday. My wish would be that he should unite with us, furnish the means of proceeding with the patent and instruments, receiving in return shares. You may mention the nature of the instrument, in which case you had better add that the methods of securing the wires, and of detecting injury done to them, have been carefully arranged to Faraday's satisfaction. The instrument I have made is a working one of full dimensions; the expenses attending a patent and experiments, even if tried on an extended scale of many miles, will not, I imagine, exceed £300. My object is to find a person or persons connected with the busy world, who could effectually bring my instrument and plans before the public, and would join me in the expense and profits. . . . . I must now acknowledge that dearest mother some months ago mentioned Mr. Walker to me at Paris, but at the time my views were different; to her, therefore, I am delighted to be able to acknowledge myself indebted for so promising a hint, and that seems to give me fairer hope of success."

On the 26th January, 1837, we find a letter from Dr. Reynolds, Professor at the Royal Institution of Liverpool, to Mr. Joseph N. Walker, in which he speaks of Mr. Cooke's "pamphlet." The letter is given *in extenso* at vol. i., p. 31. On the 28th there is a letter from Mr. J. Walker, of Liverpool, which further refers to the pamphlet: "I ought sooner to have written to tell you of my having received a parcel from Ravenfield, containing Mr. Cooke's plan for telegraphic communication. . . . . The other reason is that I wished to send Mr. Cooke's papers to two persons whom I thought most competent to give an opinion on its merits. . . . . I sent them to a Mr. Wilson, who is very clever, giving lectures on chemistry at the Mechanics' Institute, and he tells me that he is much pleased with the idea, but he does not know how the instrument can be made to point to such a variety of letters, figures, and symbols. I then sent them to Dr. Reynolds, who is curious in these matters, and well understands this department of science. From him I have not yet received an opinion. I will observe that Mr. Cooke cannot urge his plan (if it can be reduced to a useful practical form) at a better period than this. Already the Committee of the Grand Junction Line are in correspondence on the subject of a telegraphic communication along the line of railway between this and London, and Capt. Watson, the engineer, is supporting a scheme that it is thought may be adopted with effect," &c.

This was enclosed to Mr. Cooke in one from Mr. Thomas Walker, who offers to accompany him to Liverpool, and to introduce him to his brother and other gentlemen likely to forward his views. Mr. Cooke accordingly visited Liverpool, and was there introduced to Mr. Booth, of the Liverpool and Manchester Railway (the inventor of railway grease and of carriage links). He also obtained letters to Mr. Joshua Walker, of London, by whom he was subsequently introduced to Mr. Glyn and Mr. Creed, the Chairman and Secretary of the London and Birmingham Railroad.

He remained some time in Liverpool, endeavouring to obtain the adoption of his instrument on the incline of the Liverpool tunnel, which was then worked by a stationary engine and rope. His instrument, which gave 60 signals, was considered by the directors too complex for their requirements, and before simpler instruments could be made they were compelled to adopt a pneumatic telegraph, which had been previously under consideration. Nevertheless, soon after his return to London, he had, by the close of April, two simpler ones in working order. His expenditure up to this period was £381 8s. 10d., and he had not yet met Wheatstone.

We now come to a time when he took a step which affected most profoundly both his happiness and his public reputation. He called upon Professor Wheatstone, who had then recently immortalised himself by his exquisitely beautiful method of determining the velocity of the electric current, and, after a few interviews, they agreed to work together in partnership. The glory of Wheatstone's name and reputation at once overshadowed and eclipsed that of Mr. Cooke, who was regarded by the scientific world as a mere business partner (and by some as a mere "practical mechanic"*) whom Mr. Wheatstone had selected to work out his ideas and inventions. Mr. Wheatstone has been accused—with great apparent justice—of giving countenance and currency to these views,† which have become so universal among scientific writers that until the present generation has passed away, and the subject is examined anew, Mr. Cooke's name cannot hope to receive that credit to which it is justly entitled. No one is competent to speak with authority on this question till he has carefully mastered the two volumes from which we have so frequently quoted, and which contain the arbitration papers and the pamphlets of Mr. Cooke and Professor Wheatstone at full length. These, with two pamphlets issued by the Rev. T. F. Cooke,‡ and the extracts from the private correspondence which I am now giving, afford, however, ample materials for forming an opinion.

The letter of Mr. Cooke, of the 27th February, 1837, giving an account of his first interview with Professor Wheatstone, is a long

---

* *Quarterly Review*, June, 1854. † Cooke, vol. i., p. 114.

‡ "Authorship of the Practical Electric Telegraph of Great Britain." By the Rev. Thomas Fothergill Cooke, M.A. Simpkin, Marshall, & Co., London, 1868. "Invention of the Electric Telegraph: The Charge against Sir Charles Wheatstone of Tampering with the Press," &c. Simpkin, Marshall, & Co., London, 1869.

and interesting one. He says: "I tried last week an experiment upon a mile of wire, but the result was not sufficiently satisfactory to admit of my acting upon it. I had to lay out this enormous length of 1,760 yards in Burton Lane's small office in such a manner as to prevent any one part touching another; the patience required and fatigue undergone in making this arrangement were far from trivial. From Monday evening till Thursday night I was incessantly employed, and by Friday morning at ten o'clock all was obliged to be removed. Dissatisfied at the results obtained, I this morning obtained Dr. Roget's opinion, which was favourable, but uncertain. Next Dr. Faraday's, who, though speaking positively as to the general results formerly, hesitated to give an opinion as to the galvanic fluid action on a voltaic magnet at a great distance, when the question was put to him in that shape. I next tried Clark, a practical mechanician, who spoke positively in favour of my views; yet I felt less satisfied than ever, and called upon a Mr. Wheatstone, Professor of Chemistry at the London University, and repeated my queries. Imagine my satisfaction at hearing from him that he had four miles of wire in readiness, and imagine my dismay on hearing afterwards that he had been employed for months in the construction of a telegraph, and had actually invented two or three with the view of bringing them into practical use. We had a long conference, and I am to see his arrangement of wire to-morrow morning, and we are to converse upon the project of uniting our plans and following them out together. From what passed, my plan, if practicable, will, I think, have advantage over any of his, but this remains to be proved. Under all circumstances, I should be happy to have a scientific man for my coadjutor, though in that case I must sacrifice a large portion of the advantages. Yet I value his aid much more. You will know him as the person who invented the plan for ascertaining the velocity of lightning by means of a revolving mirror. But for the fear of increasing your anxiety, I should not have written till after our meeting to-morrow, though most likely several days must elapse ere anything definite can be concluded. I cannot say I have enjoyed myself much since coming to London; every moment has been anxiously occupied. I have been engaged in moving rapidly from place to place, and holding these interviews ever since ten o'clock, and it was four when I commenced writing. I do hope a few days will enable me to decide upon some plan, and either condemn my instrument or set it going. This suspense becomes too wearing. . . . . I will gladly write to dearest B. as soon as anything is settled, but at present all is doubt and uncertainty. In truth, I had given the telegraph up since Thursday evening, and only sought proofs of my being right to do so ere announcing it to you. This day's enquiries partly revives my hopes, but I am far from sanguine. The scientific men know little or nothing absolute on the subject. Wheatstone is the only man near the mark. I cannot explain the point, on which so much depends, to you on paper. Tom may, when I say that a lengthened course MAY convert *quantity* into *intensity*, in which case no magnetic quality is imparted to the iron magnet, as quantity alone produces that effect. This is a very subtle point, on which the 'Doctors' know nothing, and had not thought till I put it into their heads. You will see by this unconnected letter that I am writing in great haste, and with rather a confused head after the fatigues of the day. After dinner I shall be all right again, but at present my stomach wishes to relieve my head by working and allowing its fellow to repose. I shall not regret all this in the end if I succeed at last."

In a letter to his father he remarks: "Faraday and Roget have been very civil to me; the former begs he may know the result of my experiments, as he deems they may lead to more general important results."

At this stage Mr. Cooke had his chronometric dial telegraph, and his galvanometric instrument with six wires and three suspended needles. He had also invented his alarum. We learn from the arbitration papers that Mr. Wheatstone was not more advanced, having only a similar suspended needle telegraph and a permutating key-board with four keys, by which the circuits could be more conveniently combined and manipulated.* Neither party had made any publication of his inventions. The first to do it was Professor Wheatstone, and under somewhat peculiar circumstances. In the *Magazine of Popular Science* for March, 1837, there is a long letter from Munich, under the signature of O., dated December 23, 1836, containing, among other scientific information, an allusion to the working of Gauss and Weber's telegraph, and also an allusion to a telegraph on the same plan constructed by Steinheil; the number also contains a list of patents published, including some as late as the 25th of February: it was published some day in the first week in

* Cooke, vol. ii., p. 25, and Drawing iv.

March, about a week after Mr. Cooke's visit to Wheatstone. At the bottom of the last page, just under the notice of Gauss and Steinheil's telegraph, is a short article, included in brackets, containing a reference to some telegraphic experiments performed by Wheatstone about six or seven months previously, evidently inserted after the remainder of the articles had been completed and set in type. The information given could scarcely have come from anyone but Professor Wheatstone himself, and by his references to the paragraph he has tacitly admitted its authenticity; and it is hardly possible to doubt that, having been permitted to see the proof sheets of Gauss, Weber, and Steinheil's telegraphic experiments, and having, on the 27th of February, been visited by Mr. Cooke in reference to the same subject, he furnished the editor with the notes in question, which were inserted at the last moment in brackets. The question is only of interest because it has been made the ground of a statement that Professor Wheatstone had published an account of his experiments before he became acquainted with Mr. Cooke, which is certainly not the case.*

As we are following a strictly chronological arrangement, we will, before proceeding further, allude to a resolution which passed the House of Representatives of the United States at this time, viz., on the 3rd of February, 1837—"Resolved, that the Secretary of the Treasury be requested to report to the House at its next session upon the propriety of establishing a system of telegraphs for the United States." In compliance with this a circular was issued on the 10th March by the Hon. Levi Woodbury, asking information as to the propriety of establishing such a system; the circular alludes to communication "by cannon or otherwise, or in the night by rockets, fires, &c.," and had no special reference to electric telegraphs.†

On the 15th of April, Sydney E. Morse, a brother of Professor Morse, and, according to Dr. Hamel, the editor of a New York paper,‡ wrote an article in the *New York Observer* describing a plan for an electric telegraph which consisted of 26 wires, and was perfectly useless for practical purposes. In the appendix to Professor Morse's "Modern Telegraphy," published at Paris in 1867, at page 15, Professor Leonard Gale, who was a partner and friend of Professor Morse, and who resided in the same house, gives it in evidence that in March and

* See Cooke, vol. ii., 153; and Shaffner's "Telegraph Companion," p. 157.

† Vail's "American Electro-Magnetic Telegraph Companion," 1845, p. 67.

‡ "Hamel," p. 62.

April, 1837, the announcement of Gonon and Servill's telegraph (which proved to be a visual one) induced Professor Morse to consent to a public announcement of the existence of his invention. If this be correct, Professor Morse at this date contemplated a telegraph of 26 wires; but S. E. Morse, at page 8 in the same appendix, makes an affidavit (which was sworn in London in April, 1845) that he wrote the article in question, and though he gave a plan of 26 wires, each representing a letter, this was his own idea, and not that of his brother, Professor Morse, and that he was well acquainted with his brother's plan of using only one wire. It is difficult to reconcile these conflicting statements.

Following the order of dates, we return now to Cooke's telegraph. On the 14th March, 1837, he writes to his mother: "I have not had a moment to spare till to-day for letter-writing, even to Treeton, so must be excused for not giving earlier notice of the results of our experiments. Mr. Wheatstone called on Monday evening, and postponed our meeting at King's College till Wednesday. The result was nearly what I had anticipated, the electric fluid losing its magnetising quality in a lengthened course. An idea, however, suggested itself to Mr. Wheatstone, which I prepared to experiment on last Saturday, but again failed in producing any effect. I gave up my object for the time, and proposed explaining the nature of my discomfited instrument to the Professor. He, in return, imparted his to me. He handsomely acknowledged the advantage of mine, had it acted; his are ingenious, but not practicable. His favourite is the same as mine, made at Heidelberg, and now in one of my boxes at Berne, requiring six wires, and a very delicate arrangement. He proposed that we should meet again next Saturday, and make further experiments. For a time I felt relieved at having decided the fate of my own plan, but my mind returned to the subject with more perseverance than ever, and before three o'clock the next morning I had re-arranged my unfortunate machine under a new shape. Tom will understand when I say that I now use a true magnet of considerable power, with the poles about 4 inches apart, and suspended on the plane with its poles by a pivot (like a mariner's compass), a slender armature $4\frac{1}{2}$ inches long, covered with several hundred coils of copper wire (covered with silk). On passing through this coil a stream of galvanism, the armature becomes polarised, and is attracted by the poles of the magnet. The magnet is thus shaped: The armature is represented by the dots, one end being on this side of the north pole, its fellow on the other side of the south pole. Seen from above thus [*gives a sketch*] :—

"Pivot Z C. The ends of the armature magnetised, Z negatively, and C positively. Whenever the galvanic circuit is completed and worked, the ends are respectively attracted by the north and south poles of the magnet with a sufficient force to enable them to overcome the opposition of a feeble spring; the movement will not exceed $\frac{1}{20}$th of an inch. A lever forming part of the detent of my fan is moved by the pin projecting from P, and liberates the clockwork. Tom has seen an arrangement of this description in the Adelaide Gallery, used there merely as a toy. I have determined on making two instruments on this principle, and trying them with Mr. Wheatstone's four miles of wire. Moore has already commenced them. I have made further simplifications of the plan I showed you at Hastings, using no quicksilver triangles or springs, every part restoring itself by balance weights. The cost will not exceed £8, and may be made hereafter for about £2 each. I hope (if I can reduce the cost to this amount) to be able to introduce it into our public offices, banks, &c., and perhaps private houses, as I have a battery in view, *four inches* cube, which will continue in action for 10 days or more without cleaning (no acid used). I may get it employed on short distances on railroads at once, and experiments may enable us to use it for greater distances hereafter, when the requisite thickness of wire is better understood, and batteries in greater perfection. Thus you see I am deeper in it than ever. I am hastening to a philosophical coterie this evening, where I hope to gain further insight into the construction of the galvanic arrangements."

In this letter we have Mr. Cooke's impressions of the interview at which he and Mr. Wheatstone first compared their plans, but perhaps the most interesting part is that in which he describes his newly conceived form of electro-magnetic relay, the idea of which seems to have been suggested to him, not by anything that he saw at King's College, but by an arrangement which he and his brother had seen at the Adelaide Gallery. At any rate, he had now grasped the importance of covering his electro-magnet with "several hundred coils of fine wire," and of having a movement which would "not exceed $\frac{1}{20}$th of an inch;" and the sketch given in his letter also indicates the use of fine wire. It is evident that on this principle (using a permanent horse-shoe magnet, standing vertically, with a light bar pivoted on an axis, covered with fine wire and lying diagonally between its poles and working nearly in contact with them), he had the means of obtaining powerful attractions or repulsions, and of working at great distances. Professor Wheatstone, however, who must have seen this arrangement, does not appear to have appreciated its merits, for he afterwards says:* "Mr. Wheatstone succeeded in constructing electro-magnets possessing power sufficient for delicate movements, and which acted at very considerable distances;" and again: "His improved electro-magnets enabled him to ring alarums at very considerable distances without the intervention of the secondary circuit which was formerly employed." In his letter to Mr. Cooke of October 20th, 1840, after alluding to the causes of their estrangement, he says: "This led me to resolve to interfere with you as little as possible, and to carry on my future researches alone, and to inform you only of the results when obtained. After this resolution had been taken, I commenced a series of researches on the laws of electro-magnets, and was fortunate enough to discover the conditions which had not hitherto been the subject of enquiry, by which effects could be obtained at great distances. This rendered electro-magnetic attraction for the first time applicable in an immediate manner to telegraphic purposes." At vol. ii, p. 87, he says: "I saw and told him it could not act, because sufficient attractive power could not be imparted to an electro-magnet in a long circuit; and to convince him of this I invited him to King's College to see the repetition of the experiments on which my conclusion was founded."†

From all this it is evident that Professor Wheatstone at this time did not appreciate the importance of using fine wire, and that he had not studied Professor Henry's paper on electro-magnets in the 20th vol. of *Silliman's Journal* for January, 1831, in which he so clearly shows the advantage of using long fine wires and numerous elements for long circuits.

On the 20th March, there is a letter from Mr. Alfred King, of Liverpool, informing Mr. Cooke that the Liverpool tunnel was 2,250 yards long, and that the apparatus for the air-tube telegraph and the groove in the side of the tunnel were ready; that Mr. Booth did not anticipate there would be any objection to Mr. Cooke making a trial on the terms he proposed.

---

* Cooke, vol. i., p. 66. † See also Cooke, vol. ii., p. 93.

On the 1st April, Mr. Joseph Walker alludes to the pneumatic telegraph at Liverpool, and says the Directors were disinclined to incur any expense for experiments, but would afford every facility in their power.

On the 8th April, Mr. Cooke writes: "I have been extremely busy with enquiries and experiments on the galvanic battery. I have hit upon one important point, and found a battery upon principles likely to secure my object—of this, more hereafter. I commenced at seven yesterday morning, and never left the room till two a.m. this morning, working without intermission for 19 hours; this exceeds what I ever did when modelling. My instruments are going on famously, though slowly."

11th April, 1837.—We learn from Dr. Hamel,* and from the Smithsonian Papers,† that Professor Henry and Professor Bache visited Europe early in April, and that on the 11th April they called upon Professor Wheatstone at King's College, where "he explained to us his plan of an electro-magnetic telegraph, and, among other things, exhibited to us his method of bringing into action a second galvanic circuit: this consisted in closing the second circuit by the deflection of a needle. I informed him that I had devised another method of producing effects somewhat similar: this consisted in opening the circuit of a large quantity magnet by attracting a movable wire by a small 'intensity magnet,' or magnet wound with fine wire." We thus see the use of fine wire for magnets again thrust upon Professor Wheatstone, and yet it was long after this that he convinced himself of its importance. The mention of the secondary circuit is important, as long after this date, and after many personal conferences with Professor Henry at Princeton, Morse patented the relay circuit, and claimed to be the first inventor of it.

On the 13th April, Mr. Cooke writes to his mother: "I have left Bury Street, being in need of an extra room to try galvanic experiments, which I am conducting with all the minuteness, accuracy, and ingenuity I can summon to my aid; the result will, I flatter myself, be the most efficient battery for experimental telegraphic purposes yet brought to light. I can keep up perfect steadiness of action for a great length of time, and the moment I have done with it, a touch of the hand removes it from the action of the acid, and by exposing the plates to the

* Hamel, p. 62.

† Proceedings of the Smithsonian Institution, March 16, 1857, p. 32.

air restores them to their original activity. My instruments will be finished early next week."

On the 15th April, the letter of Professor Morse's brother about the telegraph with 26 wires appeared in the *New York Observer*, as before described.

On the 25th, Mr. Cooke writes: "Instruments progressing rapidly. I hope my next will inform you of ulterior proceedings with regard to the patent. All goes on most promisingly, particularly my galvanic arrangements, which promise great things." On the same date, there is a very long and interesting letter from Cooke to Wheatstone about their respective claims to the invention of the relay circuit, recalling the circumstances, and showing that they had both apparently thought of it at the same time independently, and by slightly different methods —Cooke, by his electro-magnetic bar, repelled by the poles of a permanent horse-shoe magnet; and Wheatstone, by a needle and pin dipping into a cup of mercury. At the close of this month Cooke had two of his simpler form of chronometric instruments working together.* On the 3rd of May he writes: "I do congratulate myself sincerely on being able to give you something like good news respecting my instruments, both of which I have had at home, and in working order since Saturday. By a better and more comprehensible method than that I explained in my letter from or after my return from Liverpool, I obtain 41 signals, and I may as well give a slight sketch of my plan, as it is now arranged, which Tom will further illustrate. My barrel has seven pins, corresponding with as many movable keys, each giving me a distinct signal."

The description of this instrument is very long, though it does not explain the interior mechanism. He gives a sketch of a revolving dial plate having fourteen letters and figures on the circumference, in seven divisions. An index moved by an electro-magnet through a short distance indicates which of the two figures is to be read, being actuated by positive and negative currents, but it is not clear how the 41 signs are to be given. It is strange that with so many different characters he does not propose to make it capable of transmitting the whole alphabet, but omits many of the letters, substituting figures and signs in their stead. Upon the whole, the instrument, the result of such long cogitation and experiment, is disappointing, and one is not surprised at Wheatstone, with his exquisite mechanical appreciation,

* Cooke, vol. i., p. 33.

criticising it as severely as he did. The principle was good, and in his hands it might at once have been made a great success; but, not being the father of it, he appears to have treated it with indifference.

On the 4th May, he writes: "My leaving town to-morrow depends on the result of a meeting I am to have with Professor Wheatstone. I had a conversation through the two telegraphs yesterday."

On the 6th May, letters pass between the two partners with respect to the payment of the cost of their new patent, which was applied for at about this date.

On the 11th May, he writes: "At the King's College I was introduced to the Messrs. Enderby, enterprising, determined men, who say there will be no difficulty in raising the capital, &c. As I may hereafter find it convenient to have my plans and improvements established by letters bearing post-mark, I shall describe what I did for Tom's immediate satisfaction, and my own ultimate security. The changes being on the old principle—not the chronometrical arrangement—they are of so promising a nature that the old discarded plan will probably take the lead in some cases. You understood the plan by which I made four wires complete three circuits, and, by reversing the same, three more. . . . . By pausing double time on any of the contacts a second signal may be represented by each wire, making the number of signals 12. By connecting two lines of wire with one pole a fresh variety of signals may be given. Now my present arrangement is this: W R (in Fig.) is Wheatstone's wire rope consisting of four wires, each covered with tarred flax, and then enclosed in a rope."

He then describes a peculiar commutating key acting on three galvanometer needles suspended on silk threads; it is described as capable of giving 60 signals, and rang an alarum. The passage about making a second signal by "pausing double time on any of the contacts, reminds one of the Morse alphabet.

On the 23rd May, he says: "I heard from Mr. Enderby yesterday that Mr. Joshua Walker seemed favourable to the undertaking. The Enderbys are very wealthy; they have a large sail-cloth and rope manufactory at Greenwich. I am going down to-morrow to see their works and arrange a method of covering our wires with rope yarn, and include them in a rope for our cross-Thames experiment. I sent the wire off yesterday. This rope, 1,500 feet in length, including 6,000 feet of wire, is to be ready by the close of the week. My two new instruments, on the Heidelberg plan, are in a great state of forwardness. Wednesday will see one finished and the other nearly so. . . . . I am making a model of Professor Wheatstone's instrument. After seeing Mr. Walker next week, I shall, if encouraged by him, proceed with a rope and wires for the London Terminus of the Birmingham Railroad, where they require a signal communication. Hoppner furnishes me with the measurement, &c. The same rope I send to Liverpool to exhibit there."

On the 27th May, his brother sends him also a long prospectus of the proposed Fire and Police telegraph.

During this month, Cooke's needle instrument was shown to the Solicitor-General, together with a pasteboard model of Wheatstone's diagram,* or hatchment dial instrument, with five vertical needles arranged in a row, and indicating by their convergence in pairs the letters above or below. This plan of suspending the needles vertically on axles, like dipping needles,† was used by Wheatstone, and was a very great advance on anything that had been done previously, though it soon simplified itself into the double needle telegraph, and eventually into the single.

On the 10th June, he says: "Enderby Brothers write that they have finished the rope. Yesterday I went to King's College to meet Professor Wheatstone and try my instruments: I had hoped to have our experiments made public to-day, but dare not till the patent is out, as one day's impatience may ruin all." He concludes by a postscript in large capital letters: "P.S.—Patent signed by His Majesty, and receiving the Great Seal this day!!! June 10th, 1837! all now is safe!"

The patent bears the date of June 12th, 1837.

The patent being completed, they were now able to exhibit their apparatus in public, and before the end of the month a bright gleam of sunshine falls upon the scene. Mr. Cooke receives from Mr. Creed, the Secretary of the London and Birmingham Railway, then in course of construction, a very friendly letter of introduction to Mr. Robert Stephenson, adding: "The Chairman would wish, when Mr. Cooke has completed his apparatus for the experiment, that Mr. Prevost should be present. It is understood that Mr. Cooke will communicate freely with you on this interesting subject."

---

* Cooke, vol. ii., p. 169.

† The Abbé Moigno and many others erroneously state that Schilling employed vertical needles. Hamel, p. 56. [The vertical needle was exhibited by Ampère in 1820 before the Academie des Sciences. See *Journal de Physique*, 1820.]

On the 2nd July, we find an account of the results of this introduction, from which we may gather some idea of the energy and enthusiasm which Mr. Cooke threw into all his labours :—

"I could not find time till now to write, every moment being engaged from six in the morning till ten at night. On Friday I saw Mr. Joshua Walker, and imparted my plan for the fire telegraph. He spoke handsomely of it, but recommended my proving the practicability of the general principles before I attempted to introduce a project involving the disturbance of the pavement. I then expressed my wish to try experiments on the railroads. 'There,' he said, 'I can at once assist you;' and within half an hour introduced me to the Chairman and Secretary of the London and Birmingham Railroad. They both entered warmly into my views, and appointed the following day for a further consideration of the subject. To shorten details, by following up every opportunity that offered itself, and urging forward my suit unceasingly, I got through all the forms, had three interviews with Mr. Stephenson, the famed engineer, and got an order for 8 cwt. of copper wire by Friday last; obtained leave to occupy a vast building on the railroad, 65 ft. by 100 ft. wide, and had as many men and all the materials I could require placed at my disposal. The order was, 'Let Mr. Cooke have everything he may require.' By strenuous exertions I succeeded in collecting the above vast quantity of wire, cleared the huge workshop of men and lumber, by the constant labour of from 30 to 40 men, and had nearly half a mile of wire arranged by Friday night. Proceeding slowly on Saturday morning—having to teach all the men employed, viz., eight carpenters, two wire workers, and eight boys, their distinct duties—we got forward more rapidly towards evening, and at five o'clock, when the men left off work, I had about four miles of wire well arranged, and hope to get all nearly done by tomorrow night. You may imagine the task when I tell you that 2,888 nails have been put up for the suspension of the wires. The labour can only be conceived by witnessing our proceedings. I am anxious to show as much activity and accuracy in my arrangements as possible, or I might proceed more leisurely. The Secretary, however, enquired from the Chairman whether I could superintend the laying down of such a communication along the line to Birmingham, should the Company hereafter determine upon it. . . . . Should my movements progress as I could wish, I hope to show my experiments to Mr. Stephenson on Thursday or Friday, when he will be in town, he having been appointed by the Directors to report thereon. My application in the first place was to try an experiment at my own expense; but, finding the parties I addressed listened to me, I finally proposed their bearing me through free of expense, which they unhesitatingly have done in the most liberal manner. You must not expect to hear from me before next Saturday, and then only the result of my experiments, as I may not hear the nature of Mr. Stephenson's report for a month."

On the 3rd July, Mr. R. Creed writes from the offices of the London and Birmingham Railway Company, which were at that time at 33, Cornhill: "Dr. Pliny Earle and Mr. Stacey are desirous of being present at the experiment which you propose trying to-morrow morning."

The letter to his mother dated Tuesday, July 4th, gives an account of this experiment: "I have completed my line of wires, extending for 13 miles, and shall have about 2½ miles more, extending to Camden Town and back, laid down as soon as I try my final experiments. I was hard at work all day yesterday, and towards evening I received a message from several of the London and Liverpool Directors, expressing their wish to see any experiments that could be tried along the line, and they would stay in town another day purposely. I promised to do all I could, and worked till ten at night, and commenced again at four this morning. All my wires were brought to a table at one end of the room, and neatly arranged over-night, but I would try no experiment till the morning, dreading lest some of my contacts should prove imperfect. Burton Lane was with me by six this morning, when I applied my battery and tried a length of two miles first—all right; then two more with the last—all right; then 8, 10, 12, and 13, with the same result. All were tried in about one minute, so that the adjusting of my instruments, and sending of messages through a total of 55 miles, required scarcely as many seconds. I only arranged the simplest of my instruments, and had all ready by twenty minutes past nine.

"I then went home and had a good wash, took one mouthful of breakfast, and got back by ten o'clock, the hour appointed. About twenty of the Directors were soon assembled (Mr. Wheatstone could not be present), so I commenced my explanations, and got through them with all the ease and coolness imaginable. I should not have been less nervous had I been explaining them only to you. I said I had hastened my preparations not to disappoint those Directors who were leaving town, but did not offer them as an example of what my telegraph could do, but to show that the current of fluid would pass through miles of wire instantaneously, &c. I commenced by putting my Heidelberg instrument in motion, which excited great interest. I then rang a bell, &c., and finally displayed the gradual decrease of galvanic energy in lengthened currents by transmitting the current first through two miles, &c., and so on to the 13th. All expressed themselves satisfied with the principle, and seemed to take the deepest interest in the experiments. One final experiment will be made next week. Mr. Stepnenson was present, and played with the instrument more than anyone else. We are to exhibit before him and a Mr. Prevost finally. We mean to try whether a machine called an 'Electro-Magnetic' cannot be made to supersede the Galvanic Battery. I have long been anxious to ascertain this point, it being part of my original plan."

The instruments used on this occasion were his suspended needle telegraph and his "mechanical" or chronometric telegraph.

About this time we find a letter from Professor Wheatstone to Mr. Cooke, without date: "As you state that some of the Railway Directors are leaving town on Wednesday, I will prepare the apparatus for two o'clock on Tuesday, though it will interfere with the preparations for my lecture. I received a letter from Mr. Stephenson that he and Bagster would come to the College to see my experiments, but they did not come.

"With respect to your application for some cash, I am sorry I cannot at present comply, &c.

"I have spent within the last week or two a good bit for my signal boards, instruments for completing the secondary circuits, alarum, new battery, &c.

"I am anxious to get my last telegraphic apparatus complete. because I am perfectly convinced it will be the only efficient and available one."

On the 9th July, Cooke writes to his father at Hastings:—

"Many thanks for your letter. I had seen the same advertisement, by a Mr. Alexander, in the *Times;* there is nothing new in the details, and they make no mention of alarums, the very pith of our patent. The result of our experiments far exceeds my anticipations. My instruments for bringing a secondary battery into action, at the distance of 14 miles, act under the influence of six plates of my battery to admiration. Professor Wheatstone had calculated upon seven plates to a mile, and I upon two. I do not think that more than one pair per two miles will be required, and those very diminutive. I turned a needle rapidly 90° with a tip of zinc amalgamated wire, and another tip of copper wire, and a little dilute acid on my finger, through a circuit of 14 miles; this wonder I have not yet shown to anyone.

"Monday night.—Mr. Stephenson and Mr. Creed have been here to-day,* and took the deepest interest in the experiments; they wish to see the effect in greater distances still, and I have orders to fit more wire, and extend along the road. I need not say that each little delay adds a grey hair to my temples, and a wrinkle to my brow."

On the 10th, Mr. Creed writes: "I have appointed Mr. Prevost and Mr. Stephenson to meet us at Euston Station at half-past eleven, to witness the trial of your experiments," which are evidently those alluded to in the postscript to his letter to Dr. Cooke.

July 19th.—We have already remarked that an allusion was made in the March number of the *Popular Science Magazine* to a telegraph by Professor Steinheil, of Munich, on Gauss and Weber's principle. It is evident that he continued to pursue the subject, for we learn from the *Comptes Rendus* of September, 1838, in a communication from Arago to the French Academy of Sciences, and from Sturgeon's *Annals of Electricity* for March and April, 1839, that on the 19th July, 1837, Steinheil had made a telegraph on a different construction, and had seven miles of it in operation.† No publication of this experiment appears, however, to have been made till 1838, when it appeared in the *Comptes Rendus*.

On the 24th July, we find a letter from Mr. Prevost, saying that in crossing the tunnel on the Camden incline to avoid a passing train, he had run against the wires, and recommended their removal to one side. This was the rope with five wires manufactured by Enderby Brothers.

We now arrive at the time of the first public exhibition of Cooke and Wheatstone's telegraph, which is thus described in a letter to Mrs. Cooke, dated July 25th:‡ "Yesterday, Mr. Stephenson

---

* This must have been Monday, July 10th.

† Cooke, vol. i., p. 11. Highton's "Electric Telegraph," p. 57. Vail, p. 179. Hamel, p. 57.

‡ This would appear to make the date the 24th, but in 1876 Sir. W. F. Cooke gave the writer the following note:—"July 25.—This experiment was performed

witnessed our experiments through 19 miles of wire, extended from Euston Square to Camden Town, and declared himself so satisfied with the result that he begged me to lay down my wires permanently between those two points on my best plan, with a view to extending the communication hereafter, if the Directors approved. He wishes, also, to have all our instruments on the most approved construction, and I have consequently put several new ones in hand. He declared himself a 'convert to our system,' and seemed quite delighted at the correspondence we carried on at so great a distance, requesting me to send the word 'Bravo!' along the line more than once. It ended by his desiring me to send an invitation to Mr. Wheatstone to join us, which he politely replied to by saying he would do himself the honour, &c. Mr. Stephenson seems to have taken our telegraphs entirely under his patronage, and a more influential one we could not desire. . . . . I have just given orders for 5,000 feet of wood to be sawn in a particular manner, with grooves for the wires, which I am going to have boiled in coal tar previously to laying down.* Our wire is all ready. A variety of notices keep appearing in the newspapers respecting electric telegraphs, but there is nothing in them. Some rumours have got abroad respecting ours, which have given rise to most of them. The one tried at Munich will not answer on a lengthened line, we having tried experiments, and proved the insufficiency of the plan."

The instrument used on this occasion was Wheatstone's diamond-shaped "hatchment" instrument, with five vertical needles,† and we have an account of what was doubtless the same experiment from information supplied by Professor Wheatstone himself to a writer in

---

in the presence of Wheatstone, Stephenson, and Sir Charles Fox. Brunel and Sir Benjamin Hawes had been there in the evening." There might have been a second and more public trial on the 25th. He also spoke of "a rope of five wires, each wire of copper, and insulated by a covering of rope, and all hung up in a bundle." About this rope, see Cooke, vol. ii., pp. 131–2. In the author's copy Mr. Cooke has written: "Sub-Thames experiment before Prince Albert. The gun was fired by a signal from King's College to Mr. Walker's Shot Tower.—W. F. C., Jan. 5, 1875."

* Specimens of this telegraph are occasionally dug up on the Camden incline, and are known among telegraphists as the "fossil telegraph."

† Cooke, vol. ii., p. 49. The Electric Telegraph Company *versus* Nott and others. Chancery Proceedings, 1846, p. 125. The evidence given in this case is very interesting, on account of the eminence of the persons who made affidavits.

the *Quarterly Review* for June, 1854 (Dr. Andrew Wynter). This article, after quoting the paragraph in the *Magazine of Popular Science* for March, 1837, says: "Following up his experiment, Professor Wheatstone worked out the arrangement of his telegraph, and, having associated himself in 1837 with Mr. Cooke, a practical mechanic who had previously devoted much time to the same subject, a patent was taken out in the June of that year. . . .

"Late in the evening of the 25th of that month, in a dingy room near the booking office at Euston Square, by the light of a flaring dip candle, which only illuminated the surrounding darkness, sat the inventor, with a beating pulse and a heart full of hope. In an equally small room at the Camden Town Station, where the wires terminated, sat Mr. Cooke, his co-patentee, and, among others, two witnesses well known to fame, Mr. Charles Fox and Mr. Stephenson. These gentlemen listened to the first word spelt by that trembling tongue of steel which will only cease to discourse with the extinction of man himself. Mr. Cooke, in his turn, touched the keys and returned the answer. 'Never did I feel such a tumultuous sensation before,' said the Professor, 'as when all alone in the still room I heard the needles click; and as I spelt the words I felt all the magnitude of the invention, now proved to be practical beyond cavil or dispute.'"

This article justly gave much offence to Mr. Cooke and his friends,* and led to the publication of two pamphlets by his brother, the Rev. Thomas Fothergill Cooke, M.A.—the one entitled "Authorship of the Practical Electric Telegraph of Great Britain" (Bath and London, 1868), containing 131 pages; and the other, "Invention of the Electric Telegraph: The Charge against Sir Charles Wheatstone of Tampering with the Press" (London, 1869).

In the latter he gives a copy of a letter from the editor of the *Quarterly Review*, Mr. W. Elwin, who says: "I did not write the article on the telegraph, but I wrote that portion relative to the merits of the respective discoveries of which Mr. Cooke complains; the author of the essay was prompted exclusively by Mr. Wheatstone."

On the 28th July, Mr. Stephenson writes, asking that his experiment may be repeated before two ladies "who have had their curiosity much excited by the accounts which I have occasionally given them of the results of your experiments." This letter by its friendly tone appears to have given Mr. Cooke much pleasure, for he says to his

* See Letter No. 29.

mother: "I received such a friendly letter from Mr. Stephenson last night, asking me to show a few experiments to two ladies staying with them who were shortly going away. I must copy a paragraph. . . . Now, is this not a nice style of letter to receive from such a man? I was much inconvenienced, but laboured late last night, and got all in readiness. Everything acted better than before, and all were delighted. Mr. Stephenson more heartily friendly than ever."

On the 14th August, we have the first indication of that jealousy as to the credit to which they were mutually entitled, which afterwards became chronic, and culminated in the arbitration and award of 1841, and the various pamphlets and publications before alluded to.

On the 14th August, Mr. Cooke's solicitor, Mr. Robert Wilson, writes to him, in evident allusion to some paragraph which had appeared in the papers: "So much time had already elapsed, it was no longer advisable to insert a correction of the erroneous paragraph. I think that the best way now would be to take the opportunity of correction which would be afforded by the next insertion of a report of experiments."*

We come now to the date on which Professor Morse's telegraph is first made public. We have already seen that on the 10th March the United States Government had issued a circular on the subject of telegraphs, and that on the 15th April a letter by S. E. Morse, a brother of Professor Morse, appeared in a New York paper, describing a Morse system of 26 wires, about which, it was stated on oath, by Professor Leonard Gale (Morse's partner, and resident in the same house with him), that it had been inserted with the express consent of Professor Morse as an announcement of the existence of his invention.† By this time, too, Cooke and Wheatstone's experiments had been widely published. Other telegraphs had been mentioned: thus, we have Gauss's, Weber's and Steinheil's, in March; Alexander's telegraph had been described at great length in the *Scotsman* newspaper of 1st July, and been copied into other papers and into the *Mechanics' Magazine* of 12th August, as alluded to in Cooke's letter to his mother

---

* Mr. Cooke writes of this period: "The invention at once became a subject of public interest; and I found that Mr. Wheatstone was talking about it everywhere in the first person singular. I remonstrated with him. I cautioned him as a friend that he was getting himself into a false position."—Cooke, vol. i., p. 8.

† Morse's "Modern Telegraphy" (Paris, 1867), Appendix, p. 15.

of July 9th. Mr. Cooke also says, on the 25th July: "A variety of notices keep appearing in the papers respecting electric telegraphy, but there is nothing in them; the one tried at Munich will not work on a lengthened circuit." And Wheatstone, on the 6th September, writes: "I read an article in the *Chronicle*, copied from the *Courier*."*

At the end of August, 1837, the *Wurzburger Zeitung*, with an account of Steinheil's doings, had reached New York and been translated in a paper there on the 1st September.† Professor Henry, also, who had called on Wheatstone in England and seen his telegraphs, had returned to America and held frequent consultations with Professor Morse at Princeton.‡

With all this excitement about telegraphs it is not surprising that Professor Morse should consider it time to make his inventions public. This he did on the 4th September, in a letter to the New York *Journal of Commerce*, saying (in allusion, perhaps, to the letter of April 15th): "You recently announced that I was preparing a short telegraphic circuit to show to my friends the operation of the telegraph; this circuit I have completed of the length of 1,700 feet, and on Saturday, the 2nd, in the presence of Professor Gale and Dr. Daubeny of the Oxford University, and several other gentlemen, I tried a preliminary experiment with the register. It recorded the intelligence sufficiently perfect to establish the practicability of the plan and the superior simplicity of my mode over any of those proposed by the Professors in Europe. No account has reached us that any of the foreign proposed electric telegraphs have as yet succeeded in transmitting intelligible communications, but it is merely asserted of the most advanced experiment (the one in London), that by means of five wires intelligence may be conveyed."§

On the 27th September, he addressed a letter in reply to the circular of the Secretary of the Treasury, stating that he had invented his telegraph five years previously, and had contracted with Mr. Vail to have an apparatus working by the 1st January, 1838. He adds: "The cost of a single copper wire, $\frac{1}{16}$ inch in diameter, for 400 miles,

---

* J. J. Fahie's "History of Electric Telegraphy to the Year 1837" (London, 1884)—a work written after years of active research—gives by far the best account of these early telegraphs.

† Dr. Hamel gives another account of this experiment. Hamel, p. 65.

‡ Hamel, p. 64.

§ Proceedings of the Smithsonian Institution, March 16th, 1857, p. 32. Hamel, p. 62.

was recently estimated in Scotland to be about £1,000 sterling." (This statement is derived from the *Scotsman* of 1st July, 1837.) He proposes to use a single wire, either in pipes or suspended on poles, and adds that, in conjunction with Professor Leonard Gale, he is about to prepare a circuit of 20 miles. He again writes on the 28th November that he had experimented with complete success, at a distance of five miles, with a battery of 57 plates, and had more recently worked ten miles with similar success.

From these letters it is sufficiently evident that the American telegraph grew out of the efforts of Messrs. Cooke and Wheatstone and other European telegraphers, and that there is no ground for that claim of priority which it has sometimes been endeavoured to set up.

On the 6th September, Mr. Wheatstone writes to Mr. Cooke from 20, Conduit Street: "Mr. Bagster is extremely anxious about the Train Telegraph, for communicating between the guard and the engineer; two wires will suffice for the communication. The best means of connection between the carriages will require some consideration, and perhaps when you return you will turn your thoughts to the subject, &c. I read a short paragraph this morning in the *Chronicle* copied from the *Courier*, which has a little annoyed me. . . . . It is full of misstatements, owing, no doubt, to the writer having received his intelligence at second or third hand. The notice states that experiments have been made on the railway under the direction of Mr. Stephenson and myself, for 25 miles, and then describes in rather an unintelligible way the original apparatus with four wires, insinuating that there is no originality in the invention, but giving me credit for the adaptation."

On the 22nd September, we find a letter from Mr. I. K. Brunel, dated "Great Western Railway, 18, Duke Street.—Mr. Brunel presents his compliments to Mr. Cooke, and, if convenient to Mr. Cooke, would be desirous of seeing him on Sunday afternoon at about two o'clock, or on Monday morning at half-past seven;" and a letter from his solicitor, Mr. Wilson, of the same date, introducing him to Sir Benjamin Hawes, M.P., who promised at once to bring the invention under the consideration of the Government; and a few days later experiments were performed before some members of the Government. On the same day Mr. Wilson writes that Mr. George Peabody had called upon him, and was desirous of taking out an American patent, and introducing the invention into the United

States; and on the 28th there is a note in Mr. Peabody's writing,* making an appointment to see the experiments at Euston Station.

About the 28th or 29th September, Mr. Cooke sends in proposals to Mr. Glyn, the Chairman of the London and Birmingham Railroad Company, for the establishment of an electric telegraph from London to Birmingham, Manchester, Liverpool, and Holyhead; the telegraph to be carried into the several exchanges, and to be open to the public generally at such uniform charges as may be agreed upon between Mr. Cooke and the Company; an Act of Parliament to be obtained, and Mr. Cooke to have a royalty of £16 per mile, and a share of the profits. This proposition was submitted by the request of the Company. This is followed on the 2nd October by a letter offering his services for the construction of a telegraph for the Railway Company. He adds: "I feel confident that Professor Wheatstone will consent to any terms I recommend, and on his return I will talk over the patent license with him."

The 6th October is the date of the *caveat* of Professor Morse's first patent.

On the 17th, Mr. Cooke's solicitor enquires whether the difficulties with Mr. Wheatstone as to priority of names† will throw obstacles in the way of concluding the agreement with Mr. Peabody for the American patent, and he naively remarks: "The English specification is published in the 'Repertory of Arts' and other works which are sent over immediately to America, and which the Americans will immediately make use of to teach them the invention, only without the ceremony of paying you for it, unless you or Professor Wheatstone choose to 'trans-Atlanticise' for a time." There is a letter from Professor Wheatstone on the subject on the same day, and on the 19th a receipt from Mr. Peabody for a parcel of papers, &c. It does not appear why no patent was ever obtained in America, but Dr. Hamel implies that the chief at the Patent Office was a friend of Professor Morse, and states that the first sentence sent along the Washington and Baltimore line in 1844 was dictated by his daughter.‡ If any undue influence were exerted in this matter, it has to be remembered that an equal injustice was done to Professor Morse, who came to Europe in 1838 to get his apparatus patented, and was refused an English patent on the ground of want of novelty.

* See Cooke, vol. ii., pp. 60, 61.

† Cooke, vol. ii., p. 61. ‡ Hamel, p. 70.

AWARD OF SIR MARC ISAMBARD BRUNEL AND PROFESSOR J. F. DANIELL.

"As the Electric Telegraph has recently attracted a considerable share of public attention, our friends, Messrs. Cooke and Wheatstone, have been put to some inconvenience, by a misunderstanding which has prevailed respecting their relative positions in connection with the invention. The following short statement of the facts has, therefore, at their request, been drawn up by us the undersigned Sir M. Isambard Brunel, Engineer of the Thames Tunnel, and Professor Daniell, of King's College, as a document which either party may at pleasure make publicly known.

"In March, 1836, Mr. Cooke, while engaged at Heidelberg in scientific pursuits, witnessed, for the first time, one of those well-known experiments on electricity, considered as a possible means of communicating intelligence, which have been tried and exhibited from time to time, during many years, by various philosophers. Struck with the vast importance of an instantaneous mode of communication, to the railways then extending themselves over Great Britain, as well as to government and general purposes, and impressed with a strong conviction that so great an object might be practically attained by means of electricity, Mr. Cooke immediately directed his attention to the adaptation of electricity to a practical system of Telegraphing; and, giving up the profession in which he was engaged, he, from that hour, devoted himself exclusively to the realisation of that object. He came to England in April, 1836, to perfect his plans and instruments. In February, 1837, while engaged in completing a set of instruments for an intended experimental application of his Telegraph to a tunnel on the Liverpool and Manchester Railway, he became acquainted, through the introduction of Dr. Roget, with Professor Wheatstone, who had for several years given much attention to the subject of transmitting intelligence by electricity, and had made several discoveries of the highest importance connected with this subject. Among these were his well-known determination of the velocity of electricity, when passing through a metal wire; his experiments, in which the deflection of magnetic needles, the decomposition of water, and other voltaic and magneto-electric effects, were produced through greater lengths of wire than had ever before been experimented upon; and his original method of converting a few wires into a considerable number of circuits, so that they might transmit the greatest number of signals, which can be transmitted by a given number of wires, by the deflection of magnetic needles.

"In May, 1837, Messrs. Cooke and Wheatstone took out a joint English patent, on a footing of equality, for their existing inventions. The terms of their partnership, which were more exactly defined and confirmed in November, 1837, by a partnership deed, vested in Mr. Cooke, as the originator of the undertaking, the exclusive management of the invention, in Great Britain, Ireland, and the Colonies, with the exclusive engineering department, as between themselves, and all the benefits arising from the laying down of the lines, and the manufacture of the instruments. As partners standing on a perfect equality, Messrs. Cooke and Wheatstone were to divide equally all proceeds arising from the granting of licenses, or from sale of the patent rights; a per-centage being first payable to Mr. Cooke, as manager. Professor Wheatstone retained an equal voice with Mr. Cooke in selecting and modifying the forms of the telegraphic instruments, and both parties pledged themselves to impart to each other, for their equal and mutual benefit, all improvements, of whatever kind, which they might become possessed of, connected with the giving of signals, or the sounding of alarums, by means of electricity. Since the formation of the partnership, the undertaking has rapidly progressed, under the constant and equally successful exertions of the parties in their distinct departments, until it has attained the character of a simple and practical system, worked out scientifically on the sure basis of actual experience.

"Whilst Mr. Cooke is entitled to stand alone, as the gentleman to whom this country is indebted for having practically introduced and carried out the Electric Telegraph as a useful undertaking, promising to be a work of national importance; and Professor Wheatstone is acknowledged as the scientific man, whose profound and successful researches had already prepared the public to receive it as a project capable of practical application; it is to the united labours of two gentlemen so well qualified for mutual assistance, that we must attribute the rapid progress which this important invention has made during the five years since they have been associated.

"M^C I^D BRUNEL.
"J. F. DANIELL

"*London, 27th April*, 1841."

LONDON, *27th April*, 1841.

"GENTLEMEN,

"We cordially acknowledge the correctness of the facts stated in the above document, and beg to express our grateful sense of the very friendly and gratifying manner in which you have recorded your opinion of our joint labours, and of the value of our invention.

"We are, Gentlemen,

"With feelings of the highest esteem,

"Your obedient Servants,

"WILLM F. COOKE.

"C. WHEATSTONE.

"Sir M. ISAMBARD BRUNEL, and

"J. F. DANIELL, Esq., Professor, &c., &c."

———o———

Printed in the United States
By Bookmasters